Climate Change Handbook

Climate Change Handbook

Jack Quinn

SYRAWOOD
PUBLISHING HOUSE
New York

Published by Syrawood Publishing House,
750 Third Avenue, 9th Floor,
New York, NY 10017, USA
www.syrawoodpublishinghouse.com

Climate Change Handbook
Jack Quinn

International Standard Book Number: 978-1-64740-039-2 (Hardback)

Cataloging-in-Publication Data

Climate change handbook / Jack Quinn.
p. cm.
Includes bibliographical references and index.
ISBN 978-1-64740-039-2
1. Climatic changes. 2. Climate change mitigation. 3. Climatology. I. Quinn, Jack.
QC903 .C55 2020
577.22--dc23

Printed in China.

Table of Contents

	Preface	VII
Chapter 1	Introduction	1
	▪ Climate	1
	▪ Climate Change	15
Chapter 2	Causes of Climate Change	37
	▪ Radiative Forcing	40
	▪ Deforestation	51
	▪ Ozone Depletion	55
	▪ Global Dimming	64
	▪ Milankovitch Cycles	70
	▪ Glacial Period	79
	▪ Orbital Forcing	81
Chapter 3	Effects of Climate Change	86
	▪ Abrupt Climate Change	88
	▪ Physical Impacts of Climate Change	91
	▪ Climate Change and Agriculture	106
	▪ Climate Change and Ecosystems	125
	▪ Fisheries and Climate Change	136
Chapter 4	Global Warming	141
	▪ Greenhouse Effect	162
	▪ Effects of Global Warming	191
Chapter 5	Climate Change Mitigation	198
	▪ Greenhouse Gas Removal	199
	▪ Carbon Dioxide Removal	200

- CCS and Climate Change Mitigation 205
- Low-carbon Diet 208
- Impacts of Climate Change Mitigation Strategies on Animal Welfare 212

Permissions

Index

Preface

This book has been written, keeping in view that students want more practical information. Thus, my aim has been to make it as comprehensive as possible for the readers. I would like to extend my thanks to my family and co-workers for their knowledge, support and encouragement all along.

The condition when the Earth's climate system results in new weather patterns that last from few decades to a million of years is known as climate change. It can be caused due to various reasons such as human activities, Earth's internal processes and external processes. Human activities are presently accelerating climate change through global warming. The effects of global warming include regional changes in precipitation, rising sea levels, expansion of deserts and frequent extreme weather events such as heat waves. Climate changes affect food security, crop yields, etc. This book provides comprehensive insights into the issue of climate change. Different approaches, evaluations, methodologies and advanced studies on climate change have been included herein. Those in search of information to further their knowledge will be greatly assisted by this book.

A brief description of the chapters is provided below for further understanding:

Chapter - Introduction

A long-term averaged weather is defined as climate. It takes into account physical factors like temperature, humidity, atmospheric pressure, wind, and precipitation, etc. Changes or disturbances in these physical factors result in climate change. This is an introductory chapter which will briefly introduce all the significant aspects of climate and climate change.

Chapter - Causes of Climate Change

There are numerous elements which are responsible for affecting the climate, resulting in climate change. These include radiative forcing, deforestation, ozone depletion, global dimming, Milankovitch cycles, glacial period, orbital forcing, etc. This chapter has been carefully written to provide an in-depth understanding of the causes of climate change.

Chapter - Effects of Climate Change

Changes in the climate pose numerous adverse impacts on the environment and human life. Some of these impacts include forest fires, tropical cyclones, droughts, global warming, etc. which severely affect agriculture, ecosystem and fisheries. This chapter closely examines these key effects of climate change to provide an extensive understanding of the subject.

Chapter - Global Warming

The overall increase in the normal temperature of Earth's atmosphere due to greenhouse effect is termed as global warming. Gases such as carbon dioxide, carbon monoxide, CFC, methane, etc. are responsible for causing greenhouse effect. The topics elaborated in this chapter will help in gaining a better perspective about global warming.

Chapter - Climate Change Mitigation

Climate change mitigation is the process of reducing severity and negative effects of climate change. Some of its aspects include greenhouse gas removal, carbon dioxide removal, CCS and climate change mitigation, low-carbon diet, etc. The diverse aspects of climate change mitigation in the current scenario have been thoroughly discussed in this chapter.

Jack Quinn

1

Introduction

A long-term averaged weather is defined as climate. It takes into account physical factors like temperature, humidity, atmospheric pressure, wind, and precipitation, etc. Changes or disturbances in these physical factors result in climate change. This is an introductory chapter which will briefly introduce all the significant aspects of climate and climate change.

CLIMATE

Climate is the weather experienced by a given location, averaged over several decades. A region's climate tells how hot or cold, wet or dry, windy or still, and cloudy or sunny it generally is. It also tells whether these conditions prevail year-round or if they change with the seasons.

Throughout history, climate has been an important factor in determining where groups of people choose to settle. While humans are resourceful enough to survive almost anywhere on the planet, most population centers are in areas where temperature and rainfall are adequate to sustain some form of agriculture. There are fewer settlements in regions of extreme dryness or cold, such as deserts or the arctic. Climate also influences how people live. It largely defines choices of architecture, clothes, food, occupation, and recreation.

Climate is determined not only by average weather conditions but also by seasonal changes in those conditions and weather extremes. Thus, for example, a climate can be described as hot and wet year-round, frigid and dry year-round, or warm and rainy in the summer and cold and dry in the winter. Certain parts of Asia and Africa have monsoon climates, meaning they are warm year-round and relatively dry for half the year yet experience heavy rainfalls over the other six months. In other regions of the world severe blizzards or hurricanes are irregular, yet not uncommon, occurrences.

The climates of the world are differentiated by many factors, including latitude (distance north or south of the equator), temperature (the degree of hotness or coldness of an environment), topography (the shape and height of land features), and distribution of land and sea. Climate is also defined by the plants that exist in a region. What's more, climate is not a fixed property of a region. The climate has changed considerably over the 4.6 billion years since Earth was formed. It continues to change today, as it will in the future.

Elements of Climate

The two criteria that are most significant in defining climate type are temperature and precipitation (water that originates in the atmosphere and falls to the ground). These criteria, in turn, are influenced by a number of atmospheric, oceanographic, and topographic factors, such as: uneven heating of Earth by the Sun; distribution of land and water; ocean currents; wind circulation patterns; the locations of high- and low-pressure systems; mountain ranges; and altitude. Elements of secondary importance to defining climate are winds, humidity, air pressure, and sunshine versus cloud cover.

Temperature

There are two temperature statistics that are relevant to climate: annual mean (average) temperature and annual temperature range. To calculate the annual mean temperature, you need a year's worth of daily mean temperatures. A daily mean is the average of a day's maximum and minimum temperatures. For instance, if a day's high temperature is 60°F (15°C) and the low is 30°F (–1°C), the average is 45°F (7°C). Then add up the average temperature for every day throughout a year and divide by 365 to get the annual mean. A given location's annual mean temperature can be compared to the world's average, 59°F (15°F), to determine the relative warmness or coldness of that location's climate.

The annual temperature range is found by subtracting the year's lowest mean monthly temperature from the year's highest mean monthly temperature. Thus, if a city averaged 80°F (27°C) in July, its hottest month, and 40°F (4°C) in January, its coldest month, it would have an annual temperature range of 40 degrees. Some places in the United States, such as San Francisco, California, have a temperature range as small as 10 degrees. The annual temperature range reveals whether or not a location experiences different seasons, which is just as important as the annual mean temperature in identifying climate type.

Standardized global temperature information has been corrected for the influences of topography by a process known as "adjustment to sea level". Temperature drops at a rate of 3.6°F for every 1,000 feet (300 meters) you ascend above sea level. Thus, when taking the temperature on a mountain at a height of 4,000 feet (1,200 meters), you would add 14.4 degrees from the reading on the thermometer to find the temperature adjusted to sea level.

If you plot out mean annual sea level temperatures throughout the world, you find bands (called isotherms) that roughly correspond to latitude. Isotherms are imaginary lines connecting areas of similar temperature. That is, temperatures are highest at the equator, decline with increasing latitudes, and are lowest at the poles. This is due to the uneven heating of Earth by the Sun. The Sun strikes Earth most directly near the equator and most indirectly (at the steepest angle) at the poles. The atmosphere, through the movement of global winds, strives to even out the distribution of heat but does not totally erase these differences.

The isotherms veer from their paths where landmasses meet the sea. This is due to the different heating and cooling behaviors of land and water (water takes longer to heat up and retains heat longer than land does). The bending of isotherms along the coasts is also caused by ocean currents, which carry warm water toward the poles and cold water toward the equator. A third cause of this temperature differential at the coasts is upwelling, the rising up of cold waters from the depths of the ocean.

Precipitation

Two measures of precipitation are relevant to determining climate: total annual precipitation and a month-by-month breakdown of when precipitation occurs. The former statistic is an indicator of the overall wetness or dryness of a region, while the latter indicates whether or not a region experiences distinct rainy and dry seasons.

Tropical rain forests, the world's wettest places, receive 60 inches (150 centimeters) or more of rain annually, while deserts and polar regions, the world's driest places, typically receive less than 10 inches (25 centimeters).

The amount of rainfall determines the type of plant and animal life a region can support. However, the distribution of precipitation throughout the year is also an important factor—particularly in relatively dry regions. For instance, if an area receives little total annual rainfall, but most comes during the summer months, that region may be suitable for agriculture. However, if the limited precipitation of a relatively dry region is spread more or less evenly throughout the year, it may not be able to sustain crops.

Wet and dry regions are scattered across the globe. There are some general trends, however, that correspond to global air circulation patterns. For instance, around the equator, where the trade winds (dominant surface winds that blow from east to west) converge and air rises, precipitation is relatively high. Precipitation is also high, about 60°, in the middle latitudes (the regions of Earth that lie between 30° and 60° latitude), where the westerlies meet the polar easterlies and air rises. Westerlies are global-scale surface winds that travel in the middle latitudes from southwest to northeast in the Northern Hemisphere (the half of Earth that lies north of the equator) and from northwest to southeast in the Southern Hemisphere (the half of Earth that lies south of the equator). Polar easterlies are cold global winds that travel across the polar regions, from the northeast to the southwest in the Northern Hemisphere and from the southeast to the northwest in the Southern Hemisphere.

In the subtropical regions around 30° latitude, where the trade winds meet the westerlies and air sinks, conditions are much drier. Most of the world's deserts are located there. The polar regions, where the polar easterlies converge and air falls, are also characterized by dryness. It should be noted that these latitudes are not fixed. As Earth makes its yearly revolution around the Sun, the amount of sunlight received at each.

Latitude changes, causing northward and southward shifts in wet and dry latitudes.

Precipitation is also connected with temperature in that warm air can hold more moisture than cold air. In the polar regions, the air holds so little moisture that precipitation is a rare event. This situation stands in contrast to the tropics, where warm air holds an abundance of water and rain is plentiful.

Another factor influencing precipitation is topography. As air travels over a mountain, orographic lifting, the upward lifting of warm air that occurs when a warm air mass travels up the side of a mountain, results in cloud formation and precipitation. Thus, the west (windward) side of a mountain range receives ample rainfall while the east (leeward) side stays relatively dry.

The positions of large-scale high and low pressure systems around the globe are a final factor influencing precipitation. These storm-producing and storm-directing zones are influential in determining the westerly and easterly (horizontal) patterns of precipitation.

Climates of the World

Although most people are familiar with certain climate types, such as tropical rain forest, desert, and tundra, no universally accepted, official system of climate designation exists. The first system to be used widely among scientific circles was the Köppen system, developed in 1918 by German meteorologist Wladimir Köppen. Köppen divided the world into five climate types, based primarily on precipitation and secondarily on temperature. They are humid tropical; arid (dry); humid subtropical; humid temperate; humid cold; and polar. Köppen then broke each of these climatic regions down into subclimates.

In the early 1930s, American climatologist C. Warren Thornthwaite proposed another system, the Thornthwaite classification. He later revised the system in 1948 and again in 1955. Thornthwaite's system is based largely on precipitation levels but also takes into account the rate at which moisture returns to the atmosphere by evaporation (the process of water changing from a liquid to a gas) and transpiration (water lost through plant leaves and stems). He developed an index using the ratio of precipitation to evaporation and added a factor called potential evapotranspiration, the total amount of water that would be produced by a combination of evaporation and transpiration, given sufficient precipitation. He translated this formula into five different climate types named for their characteristic vegetation: rain forest, forest, grasslands, steppe, and desert.

Humid Tropical Climates

Humid tropical climates, as the name implies, are warm and wet. The mean temperature for any month seldom falls below 64°F (18°C), so there is no winter. But there is plenty of rainfall in these climates. They receive on average about 60 inches (150 centimeters) of rain per year, which may be concentrated over a few months or spread throughout the entire year.

Humid tropical climates are located in belts that are 20- to 40-degrees wide on either side of the equator. They cover the largest portion of Earth of any climate zone. They account for nearly 20 percent of the land and 43 percent of the oceans, totaling 36 percent of Earth's surface. This category is subdivided into three specific climate types: rain forest, monsoon, and savanna.

Rain Forest

Rain forests are warm regions in which the rainy season usually lasts all year. Average rainfall for each month is over 2 inches (5 centimeters), and annual total rainfall usually exceeds 60 inches (150 centimeters). In some areas it may exceed 160 inches (400 centimeters) annually. On most days, heavy showers fall in the afternoon and skies clear by the evening.

Tropical rain forests exist at low elevations in regions very close to the equator. They cover the Amazon lowland of South America, portions of eastern Central America and the Caribbean Islands, the Congo River basin of Africa, the coastal area of western Africa, the East Indies from Sumatra to New Guinea, and a slice of Myanmar in southeastern Asia. Tropical rain forests also exist in isolated areas, such as the Atlantic coast of Brazil, the Pacific coast of Colombia, the Guiana highlands in northern South America, and eastern Madagascar.

Because of their proximity to the equator, rain forests experience about the same number of daylight hours year-round. Therefore, the annual temperature range is small, generally less than 6°F

(3.3°C). The biggest change in temperature comes between day and night. The average daytime temperature is around 90°F (32°C), while the average nighttime temperature is about 72°F (22°C). The high humidity and cloud cover prevent the temperature of the rain forest from rising as high as it does on summer days in the subtropics and even in some parts of the temperate latitudes.

Tropical rain forests are also called tropical wet climates. They are noted for their abundant and varied plant and animal life. Whereas rain forests cover only 7 percent of the world's land, they contain 50 percent of all species of plants and animals. This total includes over 1.5 million species of plants and animals, the greatest biodiversity in the world.

Plants grow at all different heights throughout the rain forest, from the canopy (the uppermost spreading branchy layer of a forest) to the floor. The largest plants are the towering, broadleaf evergreens, such as mahogany and kapok trees. These trees block out most of the sunlight. Near the forest edge and in clearings where sunlight penetrates, vines and shrubs predominate. The undergrowth there is so thick that it's often impossible to walk through.

The great variety of plants provides sustenance for an equally impressive array of animals. In a rain forest, you will find animals occupying every possible niche: monkeys in the treetops; colorful birds such as toucans and parrots flying overhead; lizards scaling tree trunks; frogs, snakes, and insects on the ground; and termites tunneling underground. During the day, iguanas can be seen basking in the Sun while armadillos stay in their burrows to keep cool. Several species, such as fruit bats, moths, and caimans (cousins of the crocodile), become active at night.

Monsoon

In this type of tropical climate, the year is divided into dry and rainy seasons. While monsoon regions generally receive as much total annual rainfall as do rain forests, the precipitation is concentrated in the summer months. Typically, about 120 inches (300 centimeters) of rain falls during the rainy season, causing destructive flooding. Rainfall in the spring and fall is moderate while the months of December, January, and February are dry. The official designation of a monsoon climate is one in which monthly precipitation drops below about 2 inches (6 centimeters) for one or two months of the year.

The name *monsoon* comes from the Arabic word "mausim" for "season." A monsoon is actually a seasonal reversal of wind patterns. During winter, when the monsoon area is tilted away from the Sun, the winds that sweep across the region are primarily trade winds, blowing out toward the sea. These winds bring warm, dry air to the region. When spring arrives, the winds shift and carry moisture from the Indian Ocean over the land. At this time of year, the monsoon area is almost directly beneath the Sun. The land is heated intensely, which causes the surface air to rise. This air is replaced by the moist winds, which produce heavy rains that last throughout the summer months.

While monsoon climate is limited mostly to southeast Asia, a full 50 percent of the world's population lives in that area. The monsoon region encompasses most of India; the Indus Valley of Pakistan; portions of Nepal, Bhutan, Bangladesh, Myanmar, Malaysia, Vietnam, Thailand, Laos, and the Philippines; and the southeast tip of China. Some Atlantic coastal regions of northern South America also experience monsoon climate, as do parts of central Africa.

The total annual rainfall in monsoon regions is generally sufficient to sustain forests, even through the dry period. In some areas, these forests resemble rain forests, while in other areas the forests are less dense and merge into scrublands. These lands are cultivated primarily with rice, a crop that thrives on flooding.

Animals that have evolved to survive in both wet and dry seasons live in monsoon regions. One example is the collared anteater, which eats ants in the wet season and termites in the dry season. Another example is the iguana, a reptile that can regulate temperature by changing its skin color and water loss by concentrating its urine.

Savanna

A savanna (sometimes spelled "savannah") is a relatively flat land covered with coarse grasses that are several feet tall, and scattered, stunted deciduous trees such as baobabs and acacias. Like a monsoon climate, this climate is tropical, with wet and dry seasons. Yet the dry periods in savanna regions last longer than in monsoon regions, and annual precipitation is much lower.

Savannas are located farther from the equator than are rain forests and monsoon climates. While savannas are most common in south-central and eastern Africa, they also exist in the following locations: central and eastern India, western Central America, areas of Brazil north and south of the Amazon basin, southern Florida, parts of southeast Asia, and northern Australia.

Savannas usually receive between about 30 and 60 inches (75 and 150 centimeters) of rainfall annually, much less than rain forests and monsoon regions receive. In general, the farther a savanna is from the equator, the lower its annual rainfall. The amount of rainfall in savannas is not at all predictable, either from month to month or from year to year. In a single year, a devastating flood may be followed by a severe drought. A relatively wet year may be followed by one that is quite dry.

For at least five months out of each year, the savanna rainfall typically totals less than 2 inches (6 centimeters) per month. The dry season occurs during the winter months (December, January, and February), when high pressure systems are prominent in the region. In the summer, low-pressure systems predominate, and rainfall is abundant.

In addition to wet and dry seasons, a savanna has seasonal shifts in temperature. While the mean temperature stays above about 70°F (21°C) even in the coolest months, it hovers closer to 80°F (27°C) in the warmest months (March, April, and May). In the warm months, afternoon temperatures often range between 90 and 100°F (32 and 38°C).

African savannas are teeming with wildlife. Those grasslands are regions where lions, leopards, wild dogs, and other carnivores relentlessly pursue speedy and well-camouflaged herbivores like antelopes, gazelles, and zebras.

Dry Climates

There are two types of climate considered to be dry or arid: deserts and steppes. While deserts are the drier of the two, in neither climate does annual rainfall keep pace with the rate at which water is lost due to evaporation and transpiration. Dry climates are generally found in the region between 15° and 30° latitude (although sometimes at higher latitudes) and are sometimes bordered by mountain ranges. They cover about a quarter of the world's landmass. That total is greater than for any other climate type.

A climate's designation as dry depends on both its annual rainfall and its temperature. The mean temperature must be high enough to result in the evaporation of what little rainfall there is. To illustrate this point, consider the annual rainfall received by a hot New Mexico desert. If that same amount of precipitation were to fall on cold northern Canada, it would be enough to sustain a conifer forest.

Dry climates usually receive their sparse rainfall during the summer months, when temperatures are higher and evaporation occurs more readily. However, this rainfall is often quite irregular. One location may go two years without a drop of rain, then receive four inches in a single downpour.

Desert

Deserts are the world's driest regions. These true arid climates receive less than 10 inches (25 centimeters) of rainfall a year. Often, precipitation over these lands takes the form of virga, streamers of water that evaporate into the dry air before they even reach the ground.

Deserts come in hot and cool varieties. Hot deserts, which have no cool season, are the hottest places on Earth. Their mean temperature remains above 65°F (18°C) year-round. Cool deserts, on the other hand, have an annual mean temperature below 65°F and, for at least one month out of the year, a mean temperature below 45°F (7°C).

The daily temperature in both hot and cool deserts fluctuates greatly. The reason for this fluctuation is that little humidity is available to absorb incoming sunlight during the day and virtually no cloud cover to trap the heat escaping from the surface at night. In hot deserts, daytime temperatures commonly hover between 105 and 115°F (41 and 46°C) and sometimes exceed 120°F (49°C), while at night it cools off to around 75°F (24°C). Occasionally, during the winter months, the temperature at night in hot deserts drops below freezing. In cool deserts, afternoon temperatures in the summer often reach 105°F. However, it is not uncommon for nighttime temperatures in winter to dip below 30°F (1°C).

The world's hot deserts are located in the subtropics between 15° and 30° latitude north and south, near the Tropic of Cancer and the Tropic of Capricorn. Also known as the horse latitudes, these regions contain a high-pressure belt where air from the equatorial region descends, bringing warmth and clear skies. During the summer months, the Sun is directly overhead and produces the searing heat for which these deserts are famous. Some of the deserts in this region include the Sahara of Africa; the Great Sandy Desert of Australia; and the deserts of the North American southwest and Mexico, such as the Mojave and the Sonoran.

Cool deserts are located far inland and removed from any source of water in the middle latitudes, which lie between the latitudes of 30° and 60°. Many of these are in the rain shadow of tall mountain ranges, where precipitation is often less than 5 inches (13 centimeters) annually. Examples of deserts in this category include the Atacama Desert in Chile at the base of the Andes Mountains;

the Great Basin in Nevada at the base of the Sierra Nevada; the deserts of Nepal and northern India at the base of the Himalaya Mountains; and the Gobi in Asia.

Most deserts do have some vegetation although not a great deal. This vegetation takes the form of drought-resistant plants such as cacti, which store water in their stems and have waxy coverings that diminish transpiration, and scrubby plants like the creosote bush, which has an extensive root system. There are also a variety of wildflower species that pop up quickly following a rain only to drop their seeds and whither away once the soil dries.

Animals of the desert have adapted to the limited supply of water and food in many different ways. The camel, for instance, stores water and fat in its hump, reabsorbs moisture from exhaled air, and sweats only when its body reaches extremely high temperatures. The angulate tortoise of Africa stores water in its bladder, and reptiles retain moisture due to their thick, scaly skin. Other animals, such as snakes, spend the heat of the day in a burrow, while rabbits seek out the shade of a bush. Perhaps the ultimate example of an animal adapted to desert life is the kangaroo rat. Found in deserts of the American southwest, this rodent gets all the moisture it needs from solid food and can go its entire life without drinking a drop of water.

Steppe

A steppe is a semiarid climate, meaning that it receives somewhat more rain than a desert. Steppes can be found along the edges of deserts and serve as transition zones between arid and moist areas. Annual precipitation in steppes generally measures between 8 and 20 inches (20 and 50 centimeters), all of which falls during a short wet season. This precipitation is enough to sustain short, coarse bunch grass called steppe (from which the climate gets its name), clumps of low bushes, sagebrush, and isolated trees.

Expansive steppes can be found in central Mexico, southern South America, northwest Africa, west central Asia, and surrounding the desert of central Australia. In the United States, regions with steppe climate are called prairies. These regions include the Great Plains (east of the Rocky Mountains), valleys on the northern boundaries of the Great Basin, and along the southern coasts of California.

An example of a city with a steppe climate is Denver, Colorado, which has an annual mean temperature of 50°F (10°C). Its monthly mean temperature fluctuates from 30°F (1°C) in January to 75°F (24°C) in July, giving it an annual temperature range of 45°F (7°C).

Denver receives 15 inches (38 centimeters) of precipitation yearly, most of which falls between April and July.

The large animals living in steppes are primarily migratory, grazing mammals. Examples of these include deer and antelope in the United States; gazelles and ostriches in Africa; camels in Asia; and rodents that live on or under the ground, such as prairie dogs in the United States, gerbils in Africa, and hamsters in Asia.

Subtropical Climates

Subtropical climates are those with distinct summer and winter seasons and sufficient rainfall to keep them from being classified as dry. At least eight months of the year they have mean temperatures above 50°F (10°C) and at least one month has a mean temperature below 65°F (18°C). An important criterion of a subtropical climate is that the winters are mild. The mean temperature of the coldest month always stays above 25°F (−4°C).

These climates are situated within a band running from about 25° to 40° north and south of the equator. This band begins at the edge of the tropics and extends through the warmer half of the middle latitudes, to the edge of the temperate zone. There are three distinct types of moist subtropical climates: humid subtropical, marine, and Mediterranean.

Humid Subtropical

The most outstanding characteristic of a humid subtropical climate is its hot, muggy summer. The mean temperature during the summer months is about 75° to 80°F (24 to 27°C). The nighttime temperature seldom drops below 70°F (21°C) and during the day it often rises to more than 100°F (38°C). Intense heat waves are not uncommon and sometimes last for weeks. The relative humidity is high, even during the heat of the day, which makes for even more oppressive conditions. The dew point (the temperature at which water condenses out of the air) is often above 75°F.

Winters are generally mild and, on the lower-latitude end (closest to the equator) of a humid subtropical region, it is rare for temperatures to fall below freezing. On the higher-latitude (poleward) end, frost and snow are more common, but heavy snowstorms are still rare. Conditions change rapidly in winter. One day's warm, sunny weather can give way to the next day's cold rain.

Rainfall is plentiful throughout the year, totaling between 30 and 65 inches (about 75 and 165 centimeters) annually. Afternoon thunderstorms are common in the summer, as is precipitation from the edge of tropical storms. Much of the rain and snow in winter is associated with eastward-moving storm fronts.

This type of climate is found along the east coasts of most continents, between 25° and 37° latitude. It covers the southeastern United States, eastern China, southern Japan, southeastern South America (Uruguay, Paraguay, and southern Brazil), and the southeastern coasts of Africa and Australia. These locations are mostly on the western sides of subtropical high-pressure systems, placing them in the path of maritime tropical air that originates near the equator and is directed toward the poles.

Since the growing season lasts from eight months to a year, depending on location, this climate is well suited for many types of agriculture. The natural plant cover consists of thick deciduous and coniferous woodlands at the eastern edge of these zones, graduating to prairies on the western edge.

There is a minor subset of humid subtropical climate, examples of which include Portugal and the Canadian Pacific coast, that have mild, rainy winters, and cool, dry, short summers. During winter in these places, the ocean winds keep the air warmer than it is farther inland. The average temperature for the warmest months in these regions never exceeds 72°F (22°C) and three times as much rain falls during the winter as during the summer.

Marine

Marine climate, also known as Marine West Coast climate, features cool summers and mild winters, yet the summers are longer than they are in the humid subtropical climate of Pacific Canada. Precipitation is plentiful year-round in marine climates, and there is a relatively small annual temperature range. Some parts of the world with a marine climate are the west coasts of northern California, Washington, and Oregon; the British Isles; New Zealand; the west coast of France; and the southeast coast of South Africa.

This climate is characterized by the presence of low clouds, fog, and drizzle for much of the year, as well as regular rainfall and occasional snowfall. The temperatures are generally high enough that if snow does fall, it melts quickly. These conditions, plus relatively stable temperatures year-round, are largely the result of the prevailing westerly winds that bring ocean air into these regions.

On the U.S. northwest coast, rainfall is higher in the winter than in the summer. The reason for this pattern is the subtropical Pacific high-pressure system that directs storms away from the coast in the summer. It is common to find thick stands of coniferous and deciduous trees in marine climates.

Mediterranean

The primary factor that distinguishes Mediterranean climates from other subtropical climates is that it has dry summers. In areas with a Mediterranean climate, there are four to six months each year with little or no rainfall. At least three times as much rain falls in the wettest (winter) month as in the driest (summer) month. Total rainfall for the year is between about 15 and 35 inches (38 and 89 centimeters).

The summer dryness is caused by subtropical high pressure systems located over the water, which produce sinking air and direct storms poleward, away from the coasts. The position of these highs shifts towards the equator in the summer, allowing mid-latitude storms (with plenty of rain) to enter the region.

The summer high temperatures are lowest at points nearest to the water's edge. This effect is caused by upwelling, the churning up of cold water from the depths of the ocean, which chills the coastal air. As one moves inland, the summers are hotter (and the winters are colder, too).

The largest region with a Mediterranean climate consists of the coastal lands surrounding the Mediterranean Sea. Specifically, this region includes the Iberian Peninsula (Spain and Portugal), southern France, most of Italy, western Serbia and Montenegro, Greece, southern Bulgaria, the coasts of Turkey and Israel, and the Mediterranean coast of northern Africa. This region's reputation for warm, sunny summers attracts crowds of tourists from around the world each year.

Other parts of the world with a Mediterranean climate include the central and southern California coast, the southwestern tip of Africa, the southeastern and southwestern edges of Australia, and a small area on the west coast of central Chile.

The irregular rainfall and preponderance of summer wildfires means that the only types of vegetation that can survive in this climate are shrubs, dwarf trees, and grasses. The shrubs have extensive root systems, enabling them to gather what little moisture exists in the soil. The leaves of Mediterranean plants also have moisture-conserving designs. For instance, some leaves have hard, waxy surfaces; some are hairy; and others curl up when it's dry. The olive tree has gray-colored leaves, which reflect sunlight.

Mediterranean vegetation is also adapted to periodic wildfires. After a fire, many will resprout from their roots. Some plant species, such as the knobcone pine, actually depend on fire for their propagation. The cones of these trees open and release seeds only in the extreme heat of a fire. If there is any doubt about the ability of scrubland vegetation to withstand wildfires, one only has to witness the rapid growth that follows in the wake of the burn.

The native animal inhabitants of these lands are deer, rabbits, wolves, and numerous rodents such as squirrels, voles, and mice. In some regions, like California and southern Europe, cows, goats, sheep, and other domesticated grazing animals are now the dominant species.

Temperate Climates

Temperate climates are those that have four distinct seasons. The criterion that differentiates them from subtropical (the next warmest) climates is that for a least one month of the year the mean temperature in a temperate climate is below 25°F (−4°C). Temperate climates can be distinguished from subpolar (the next coldest) climates in that they have at least one month in the year in which the mean temperature is above 70°F (21°C).

All temperate areas have warm, humid summers and cold, snowy winters, although the average monthly temperatures and precipitation levels vary considerably from one place to another. Total annual precipitation ranges from 20 to 40 inches (50 to 100 centimeters) and is steady throughout the year.

Temperate climates are situated mostly between 40° to 60° latitude, primarily in the Northern Hemisphere. They cover the northeastern United States, southern Canada, and parts of central Europe and central Asia. In the Southern Hemisphere, where there is very little land mass south of the 40th parallel, temperate climates exist only in the southeastern portion of Chile, the southern tip of Australia, and New Zealand.

Toward the lower-latitude end of a zone of temperate climate, summers are hotter and more humid than they are in the higher latitudes. High temperatures in the summer are often over 90°F (32°C) in the lower latitudes, and there are five or six months of the year during which temperatures never dip below freezing. This means that the growing season is long enough to cultivate a variety of crops. In those same locations, the winter is cold and blustery with significant snowfall.

In the colder temperate areas, winter is harsher and longer, summer is milder and shorter, and spring and fall are merely quick transitions between winter and summer. Typically, there are only three to five months a year with no frost. During the summer, high temperatures rarely exceed 90°F (32°C) while in the winter the temperature may drop below −22°F (−30°C) and can remain below zero degrees for weeks on end.

The most common type of vegetation found in temperate areas is the deciduous forest. A deciduous forest contains trees such as oak, hickory, maple, beech, ash, and birch. In some areas there are also stands of conifers (evergreens) such as pine, spruce, and fir. In the autumn, a chemical transformation occurs within the leaves of deciduous trees, causing them to turn brilliant shades of red, orange, and yellow. In the spring, before the thick blanket of leaves has fully emerged, wildflowers dot the forest floor.

A great variety of birds, mammals, reptiles, and insects inhabit temperate areas. While many species of birds and some mammals migrate to warmer regions for the winter, other animals have developed methods of surviving through the cold months. Squirrels, for instance, store enough food in the summer and fall to last through the winter. Turtles find a deep, muddy burrow in which to hibernate. With many insect species, the adults die before the onset of winter, and their eggs hatch the following spring.

Subpolar Climate

The subpolar climate zone is a cold, northern climate, found generally between 50 and 70° north latitude. It is dominated by a nearly continuous coniferous forest called the "taiga" or "boreal forest." Specifically, this region includes Alaska, central and northern Canada (except for the northernmost fringe), northern Europe (except for the north of Scandinavia), and northern Asia (except for the northern and southern fringes of Siberia). There is no subpolar land in the Southern Hemisphere because, except for the very southernmost tip of South America, no land exists at these latitudes.

Subpolar climate is characterized by long, harsh winters and short, cool summers, with a wide range of temperature variation between the coldest and warmest months. The factor that distinguishes subpolar climates from neighboring, colder polar climates is that subpolar climates have one to three months in which the mean temperature is above 50°F (10°C). There is no limit, however, defining the cold end of the subpolar temperature range. In northern Siberia, for instance, the coldest monthly average temperature sometimes drops below −35°F (−37°C).

The cold air of subpolar regions retains little moisture. Thus there is little cloud formation and hence, light precipitation. Most subpolar lands receive less than 20 inches (50 centimeters) of precipitation annually. While snowfall is light, the air is so cold that once the snow falls, it remains for months. Because of the low temperatures, large tracts of subpolar land contain permafrost, a layer of subterranean soil that remains frozen year-round.

The short growing season rules out agriculture in the region. The main type of vegetation is conifer trees, such as fir, spruce, larch, and pine, plus a few varieties of deciduous trees, such as birch and alder. Conifers are well-suited to this environment. Their tall, narrow, conical shape encourages the snow to slide off them, and their needles don't require much moisture to grow.

Large tracts of subpolar land are covered with highly acidic wetlands called bogs. Bogs are comprised primarily of sphagnum moss, hardy grasses, insectivorous plants, pondweeds, water lilies, and algae, as well as a smattering of flowering plants such as blueberry and azalea. When the moss dies, it accumulates as peat, which in some bogs forms a bouncy, carpetlike layer several feet thick.

The mammals of the boreal forest include the lynx and various members of the weasel family (wolverine, fisher, pine martin, mink, ermine, and sable). These animals prey on snowshoe hares, red squirrels, lemmings, and voles. Beaver are abundant, and their dam-building habit is an important part of forest succession. Large mammals of the boreal forest include moose and elk. Insect-eating birds such as wood warblers are migratory, but seed-eating birds (finches and sparrows, evening grosbeak, pine siskin, and red crossbill) and omnivores (ravens) may be year-round residents. All of these birds may move south if food is scarce. Mosquitoes are abundant in the spring.

Polar Climate

The polar climate, which covers the extreme northern and southern portions of Earth, is the coldest in the world. The warmest month in these regions has an average temperature below 50°F (10°C). It is so cold near the poles because the sunlight strikes the surface there at a very steep angle and remains low in the sky even in the summer. This condition is in stark contrast to the world's hottest regions, where the Sun sometimes shines directly overhead.

The polar area extends, on its warmest boundary, to between 60° and 70° latitude. The coldest points are at 90° latitude, capping the North and South Poles. Lands with polar climates include: the northern reaches of North America, Europe, and Asia; all of Greenland; the northern half of Iceland; and Antarctica.

Polar regions receive less than 8 inches (20 centimeters) of precipitation annually, which is equivalent to that received by a desert. In a desert the air is warm enough for evaporation to exceed precipitation. In polar regions, however, the air is so cold that the meager rate of precipitation is still greater than the rate of evaporation, and the ground never dries out.

The polar climate is divided into two categories: tundra, which experiences a yearly thaw, and arctic, which remains a permanently frozen region.

Tundra

Tundra comprises the majority of the polar land in the Northern Hemisphere, with the arctic being restricted to areas close to the North Pole. Due to the virtual absence of land between 50° and 70° latitude south, the only tundra in the Southern Hemisphere lies in narrow strips on the coasts of Antarctica.

In the tundra, the mean temperature of the warmest month is between 30 and 50°F (−1 and 10°C). While the tundra summers are cool, the winters are bitterly cold. An example of a tundra climate is Barter Island, Alaska, at 70° north latitude. The coldest month of the year at Barter Island is February, with an average monthly temperature of −20°F (−29°C). It has three months in which the average temperature is above freezing: in June the temperature reaches 35°F (2°C); and in July and August, 40°F (4°C). Throughout an average year, Barter Island receives about 7 inches (18 centimeters) of precipitation.

Beneath the tundra surface lies a layer of permafrost, frozen subterranean soil, hundreds of yards deep. In the summer the top few feet of soil thaw. Since the moisture can't penetrate the frozen ground below, the surface becomes muddy and swampy. The growing season is so short that only a limited number of small plant species dot the landscape. These include mosses, lichens, sedges, and dwarf trees. The trees are species of willow and birch; in warmer climates they grow several feet tall but in the tundra creep along the ground and are only a few inches tall.

Arctic

The arctic is a barren land of snow and ice located around the poles in both hemispheres. There the average temperatures for every month are below freezing. In the Northern Hemisphere arctic lands include Ellesmere Island and the northern section of Baffin Island (these islands are part of the Canadian Archipelago) and the interior of Greenland, plus the seas between these lands. In the Southern Hemisphere it includes nearly the whole of Antarctica.

While temperatures vary greatly throughout the arctic, the coldest weather is found in Antarctica. Earth's lowest temperature, −128.6°F (−89°C), was recorded on July 26, 1983, at the Soviet research station Vostok in Antarctica. The station is located at 78° south latitude with an elevation of 11,400 feet (3,475 meters) above sea level. During Vostok's warmest month, December, the average temperature is −30°F (−34°C) and during its coldest month, August, the average temperature is −90°F (−68°C). The warmest temperature ever recorded there was −5.8°F (−21°C).

Antarctica is warmest on its outskirts and coldest in its interior. In the summer months, average temperatures range from 14 to 20°F (−10 to −7°C) near the seas and from about −5 to −20°F (−21 to −29°C) near the center of the continent. In the coldest months the outer areas have average temperatures about −20°F, while average temperatures in the interior are about −95°F (−71°C).

Conditions in the arctic region of the Northern Hemisphere are not much more hospitable. Average temperatures there are about −45°F (−43°C) in January, the coldest month, and about 10°F (−12°C) in July, the warmest month. The coldest temperature ever recorded in Greenland was −87°F (−66°C) on January 9, 1954, at Northice, which is at a latitude of 78° north and an altitude of about 7,685 feet (2,340 meters). Farther south, at a latitude of 71° north, is Eismitte (which means "middle of the ice"). At this location, about 9,940 feet (3,030 meters) above sea level, the average temperature in February, the coldest month, is about −55°F (−48°C) and the average temperature in July, the warmest month, is about 10°F (−12°C).

Mountain Climates

There is no inclusive definition of a mountain climate, also called a highland climate. The reason is that a mountain climate encompasses a whole series of climate types that are found at various points on the way up a mountain. The primary reason that conditions change as you ascend a mountain is that temperature decreases with altitude. Air pressure, humidity, precipitation, and winds also change with altitude. The climates on a mountain form pockets within the general climate of the surrounding region.

Mountain climates can resemble climates found at higher latitudes. In general, each 1,000-foot (300-meter) increase in elevation will produce a change in climate similar to traveling about 185 miles (298 kilometers) towards one of Earth's poles. For every 1,000 feet (300 meters) gained in altitude, the temperature falls about 3.6°F (2.0°C). However, mountain climates are not exact matches to climate zones at higher latitudes because many factors are different. For example, the effects of global atmospheric circulation, extratropical cyclones, and other weather systems are not as significant in mountain climates.

A 13,000-foot-tall (3,960-meter-tall) mountain in the Sierra Nevada range, for example, encompasses five different climate types. A semiarid grassland surrounds the base of this mountain.

Along the "foothills," the first 2,000 feet (610 meters), the grassland is gradually replaced by Mediterranean scrubland. From 3,000 to 6,000 feet (915 to 1,825 meters), there is the "montane zone," the climate resembles that of a northern temperate zone, with thick conifer forests. Above 6,000 feet, in the "subalpine zone," thinner stands of stunted conifers resemble a subpolar climate. This zone gives way to tundra at 9,000 feet (2,740 meters), in the "alpine zone." Finally, the mountain is capped with a permanent expanse of snow and ice.

Another notable feature of mountain climates is the uneven distribution of precipitation across a mountain or mountain range, known as rain shadow effect. Mountains have a relatively wet side, called the windward side, and a relatively dry side, called the leeward side. The windward side is the one encountered by approaching air masses. The windward side is on western slopes in the Northern Hemisphere and on eastern slopes in the Southern Hemisphere. The air rides upward along the mountainside, and as it rises, it cools. Once air reaches the dew point, condensation (the process by which water changes from gas to liquid) occurs and clouds form.

The animals that inhabit mountains are, for the most part, migratory. However, they do not follow typical north-south migratory paths. Rather they migrate up and down the mountain. They climb to higher elevations in the summer, when food is more abundant, and escape the harsh alpine winters by climbing down to warmer elevations. Among the wide variety of mountain animals are deer, bears, chipmunks, mountain goats, marmots, and many species of birds. Some animals are specially adapted to survive in the oxygen-thin air found at high altitudes. One such animal is the guanaco, a relative of the llama, which has an unusually large heart and lungs.

Mountain climates are scattered throughout North America, South America, Africa, Europe, Asia, and the islands north of Australia. Examples of these climates include the Sierra Nevada and the Rocky Mountain ranges in the United States and Canada, the Sierra Madre Range in Mexico and Central America, the Andes in Chile and Peru, the Alps in Europe, and the Himalayas in southern Asia.

CLIMATE CHANGE

Climate change is the periodic modification of Earth's climate brought about as a result of changes in the atmosphere as well as interactions between the atmosphere and various other geologic, chemical, biological, and geographic factors within the Earth system.

The atmosphere is a dynamic fluid that is continually in motion. Both its physical properties and its rate and direction of motion are influenced by a variety of factors, including solar radiation, the geographic position of continents, ocean currents, the location and orientation of mountain ranges, atmospheric chemistry, and vegetation growing on the land surface. All these factors change through time. Some factors, such as the distribution of heat within the oceans, atmospheric chemistry, and surface vegetation, change at very short timescales. Others, such as the position of continents and the location and height of mountain ranges, change over very long timescales. Therefore, climate, which results from the physical properties and motion of the atmosphere, varies at every conceivable timescale.

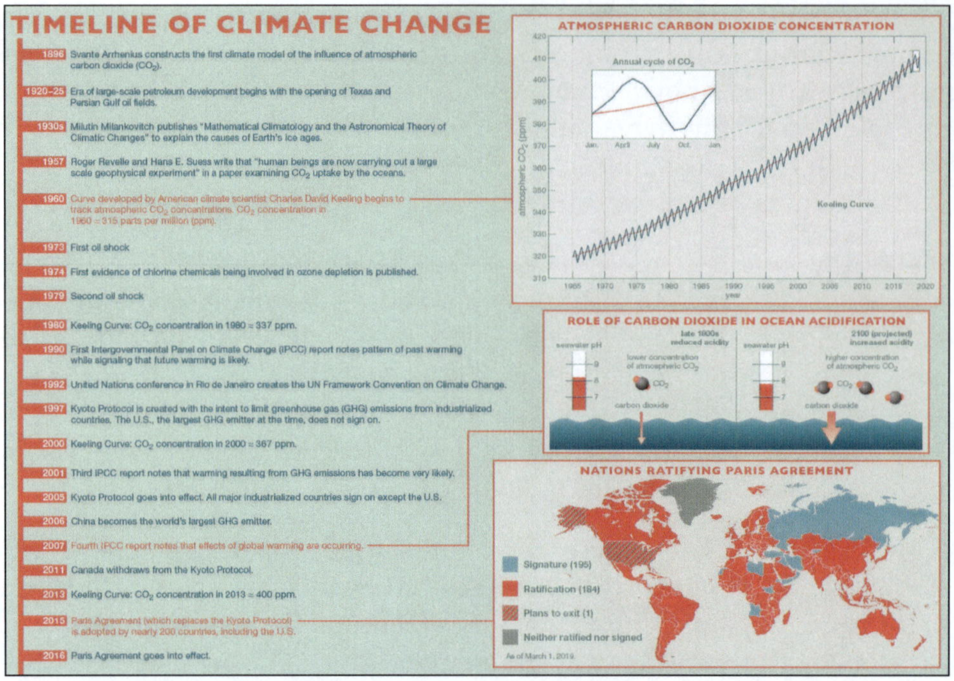

Climate change: A timeline of important developments in climate change.

Climate is often defined loosely as the average weather at a particular place, incorporating such features as temperature, precipitation, humidity, and windiness. A more specific definition would state that climate is the mean state and variability of these features over some extended time period. Both definitions acknowledge that the weather is always changing, owing to instabilities in the atmosphere. And as weather varies from day to day, so too does climate vary, from daily day-and-night cycles up to periods of geologic time hundreds of millions of years long. In a very real sense, *climate variation* is a redundant expression—climate is always varying. No two years are exactly alike, nor are any two decades, any two centuries, or any two millennia.

Earth System

Iceberg Tourist boat in front of a massive iceberg near the coast of Greenland.

The atmosphere is influenced by and linked to other features of Earth, including oceans, ice masses (glaciers and sea ice), land surfaces, and vegetation. Together, they make up an integrated Earth system, in which all components interact with and influence one another in often complex ways. For instance, climate influences the distribution of vegetation on Earth's surface (e.g., deserts exist in arid regions, forests in humid regions), but vegetation in turn influences climate by reflecting radiant energy back into the atmosphere, transferring water (and latent heat) from soil to the atmosphere, and influencing the horizontal movement of air across the land surface.

Turkmenistan Drought-resistant plants grow in the Repetek Preserve in the southeastern Karakum Desert, Turkmenistan.

Deciduous forest in fall coloration, Wasatch Mountains, Utah.

Earth scientists and atmospheric scientists are still seeking a full understanding of the complex feedbacks and interactions among the various components of the Earth system. This effort is being facilitated by the development of an interdisciplinary science called Earth system science. Earth system science is composed of a wide range of disciplines, including climatology (the study of the atmosphere), geology (the study of Earth's surface and underground processes), ecology (the study of how Earth's organisms relate to one another and their environment), oceanography (the study of Earth's oceans), glaciology (the study of Earth's ice masses), and even the social sciences (the study of human behaviour in its social and cultural aspects).

A full understanding of the Earth system requires knowledge of how the system and its components have changed through time. The pursuit of this understanding has led to development of Earth system history, an interdisciplinary science that includes not only the contributions of Earth system scientists but also paleontologists (who study the life of past geologic periods), paleoclimatologists (who study past climates), paleoecologists (who study past environments and ecosystems), paleoceanographers (who study the history of the oceans), and other scientists concerned with Earth history. Because different components of the Earth system change at different rates and are relevant at different timescales, Earth system history is a diverse and complex science. Students of Earth system history are not just concerned with documenting what has happened; they also view the past as a series of experiments in which solar radiation, ocean currents, continental configurations, atmospheric chemistry, and other important features have varied. These experiments provide opportunities to learn the relative influences of and interactions between various components of the Earth system. Studies of Earth system history also specify the full array of states the system has experienced in the past and those the system is capable of experiencing in the future.

Undoubtedly, people have always been aware of climatic variation at the relatively short timescales of seasons, years, and decades. Biblical scripture and other early documents refer to droughts, floods, periods of severe cold, and other climatic events. Nevertheless, a full appreciation of the nature and magnitude of climatic change did not come about until the late 18th and early 19th centuries, a time when the widespread recognition of the deep antiquity of Earth occurred. Naturalists of this time, including Scottish geologist Charles Lyell, Swiss-born naturalist and geologist Louis Agassiz, English naturalist Charles Darwin, American botanist Asa Gray, and Welsh naturalist Alfred Russel Wallace, came to recognize geologic and biogeographic evidence that made sense only in the light of past climates radically different from those prevailing today.

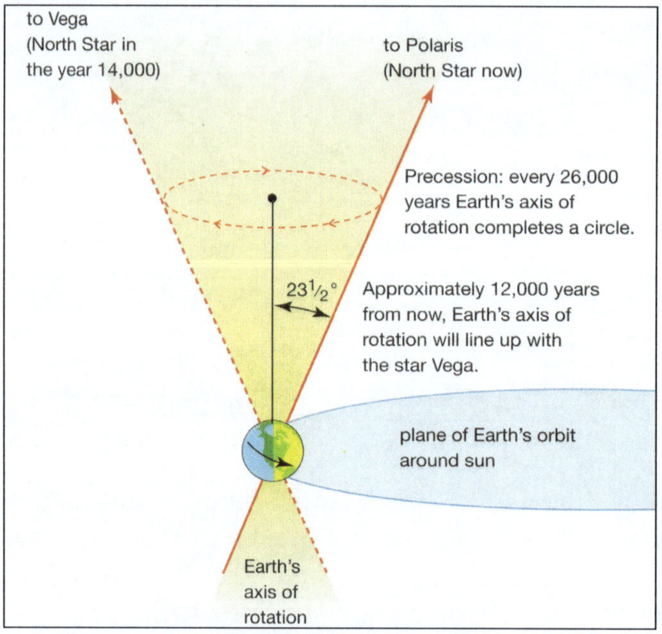

The precession of Earth's axis.

Geologists and paleontologists in the 19th and early 20th centuries uncovered evidence of massive climatic changes taking place before the Pleistocene—that is, before some 2.6 million years ago. For example, red beds indicated aridity in regions that are now humid (e.g., England and New England), whereas fossils of coal-swamp plants and reef corals indicated that tropical climates once occurred at present-day high latitudes in both Europe and North America. Since the late 20th century the development of advanced technologies for dating rocks, together with geochemical techniques and other analytical tools, have revolutionized the understanding of early Earth system history.

The occurrence of multiple epochs in recent Earth history during which continental glaciers, developed at high latitudes, penetrated into northern Europe and eastern North America was recognized by scientists by the late 19th century. Scottish geologist James Croll proposed that recurring variations in orbital eccentricity (the deviation of Earth's orbit from a perfectly circular path) were responsible for alternating glacial and interglacial periods. Croll's controversial idea was taken up by Serbian mathematician and astronomer Milutin Milankovitch in the early 20th century. Milankovitch proposed that the mechanism that brought about periods of glaciation was driven by cyclic changes in eccentricity as well as two other orbital parameters: precession (a change in the directional focus of Earth's axis of rotation) and axial tilt (a change in the inclination of Earth's axis with respect to the plane of its orbit around the Sun). Orbital variation is now recognized as an important driver of climatic variation throughout Earth's history.

Evidence for Climate Change

All historical sciences share a problem: As they probe farther back in time, they become more reliant on fragmentary and indirect evidence. Earth system history is no exception. High-quality instrumental records spanning the past century exist for most parts of the world, but the records become sparse in the 19th century, and few records predate the late 18th century. Other historical documents, including ship's logs, diaries, court and church records, and tax rolls, can sometimes be used. Within strict geographic contexts, these sources can provide information on frosts, droughts, floods, sea ice, the dates of monsoons, and other climatic features—in some cases up to several hundred years ago.

Fortunately, climatic change also leaves a variety of signatures in the natural world. Climate influences the growth of trees and corals, the abundance and geographic distribution of plant and animal species, the chemistry of oceans and lakes, the accumulation of ice in cold regions, and the erosion and deposition of materials on Earth's surface. Paleoclimatologists study the traces of these effects, devising clever and subtle ways to obtain information about past climates. Most of the evidence of past climatic change is circumstantial, so paleoclimatology involves a great deal of investigative work. Wherever possible, paleoclimatologists try to use multiple lines of evidence to cross-check their conclusions. They are frequently confronted with conflicting evidence, but this, as in other sciences, usually leads to an enhanced understanding of the Earth system and its complex history. New sources of data, analytical tools, and instruments are becoming available, and the field is moving quickly. Revolutionary changes in the understanding of Earth's climate history have occurred since the 1990s, and coming decades will bring many new insights and interpretations.

Ongoing climatic changes are being monitored by networks of sensors in space, on the land surface, and both on and below the surface of the world's oceans. Climatic changes of the past 200–300 years, especially since the early 1900s, are documented by instrumental records and other archives. These written documents and records provide information about climate change in some locations for the past few hundred years. Some very rare records date back over 1,000 years. Researchers studying climatic changes predating the instrumental record rely increasingly on natural archives, which are biological or geologic processes that record some aspect of past climate. These natural archives, often referred to as proxy evidence, are extraordinarily diverse; they include, but are not limited to, fossil records of past plant and animal distributions, sedimentary and geochemical indicators of former conditions of oceans and continents, and land surface features characteristic of past climates. Paleoclimatologists study these natural archives by collecting cores, or cylindrical samples, of sediments from lakes, bogs, and oceans; by studying surface features and geological strata; by examining tree ring patterns from cores or sections of living and dead trees; by drilling into marine corals and cave stalagmites; by drilling into the ice sheets of Antarctica and Greenland and the high-elevation glaciers of the Plateau of Tibet, the Andes, and other montane regions; and by a wide variety of other means. Techniques for extracting paleoclimatic information are continually being developed and refined, and new kinds of natural archives are being recognized and exploited.

Climate Change within a Human Life Span

Regardless of their locations on the planet, all humans experience climate variability and change within their lifetimes. The most familiar and predictable phenomena are the seasonal cycles, to which people adjust their clothing, outdoor activities, thermostats, and agricultural practices. However, no two summers or winters are exactly alike in the same place; some are warmer, wetter, or stormier than others. This interannual variation in climate is partly responsible for year-to-year variations in fuel prices, crop yields, road maintenance budgets, and wildfire hazards. Single-year, precipitation-driven floods can cause severe economic damage, such as those of the upper Mississippi River drainage basin during the summer of 1993, and loss of life, such as those that devastated much of Bangladesh in the summer of 1998. Similar damage and loss of life can also occur as the result of wildfires, severe storms, hurricanes, heat waves, and other climate-related events.

Abandoned farmstead showing the effects of wind erosion in the Dust Bowl, Texas county, Okla.

Climate variation and change may also occur over longer periods, such as decades. Some locations experience multiple years of drought, floods, or other harsh conditions. Such decadal variation of climate poses challenges to human activities and planning. For example, multiyear droughts can disrupt water supplies, induce crop failures, and cause economic and social dislocation, as in the case of the Dust Bowl droughts in the midcontinent of North America during the 1930s. Multiyear droughts may even cause widespread starvation, as in the Sahel drought that occurred in northern Africa during the 1970s and '80s.

Seasonal Variation

Every place on Earth experiences seasonal variation in climate (though the shift can be slight in some tropical regions). This cyclic variation is driven by seasonal changes in the supply of solar radiation to Earth's atmosphere and surface. Earth's orbit around the Sun is elliptical; it is closer to the Sun (147 million km; about 91 million miles) near the winter solstice and farther from the Sun (152 million km; about 94 million miles) near the summer solstice in the Northern Hemisphere. Furthermore, Earth's axis of rotation occurs at an oblique angle (23.5°) with respect to its orbit. Thus, each hemisphere is tilted away from the Sun during its winter period and toward the Sun in its summer period. When a hemisphere is tilted away from the Sun, it receives less solar radiation than the opposite hemisphere, which at that time is pointed toward the Sun. Thus, despite the closer proximity of the Sun at the winter solstice, the Northern Hemisphere receives less solar radiation during the winter than it does during the summer. Also as a consequence of the tilt, when the Northern Hemisphere experiences winter, the Southern Hemisphere experiences summer.

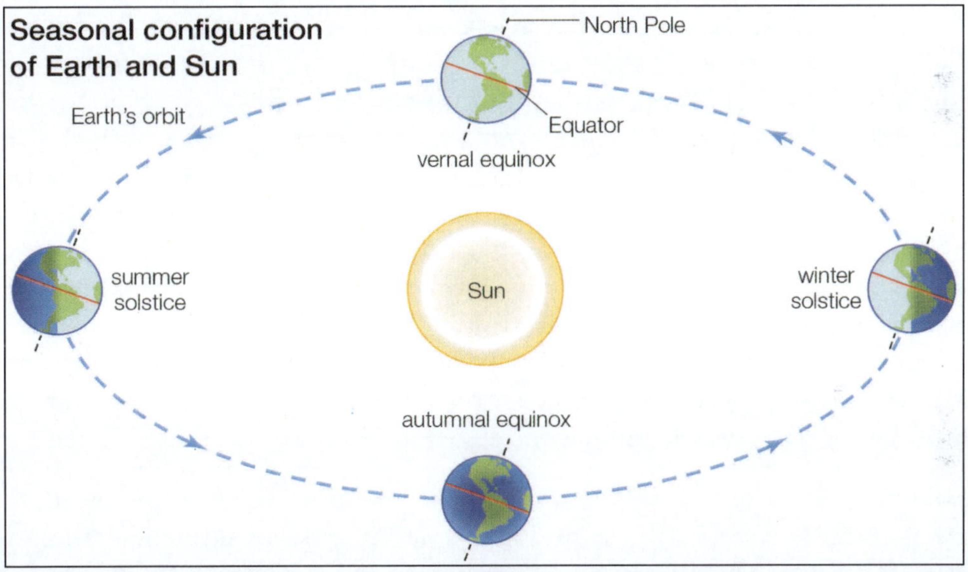

A diagram shows the position of Earth at the beginning of each season in the Northern Hemisphere.

Earth's climate system is driven by solar radiation; seasonal differences in climate ultimately result from the seasonal changes in Earth's orbit. The circulation of air in the atmosphere and water in the oceans responds to seasonal variations of available energy from the Sun. Specific seasonal changes in climate occurring at any given location on Earth's surface largely result from the transfer of energy from atmospheric and oceanic circulation. Differences in surface heating taking place between summer and winter cause storm tracks and pressure centres to shift position and strength. These heating differences also drive seasonal changes in cloudiness, precipitation, and wind.

Seasonal responses of the biosphere (especially vegetation) and cryosphere (glaciers, sea ice, snowfields) also feed into atmospheric circulation and climate. Leaf fall by deciduous trees as they go into winter dormancy increases the albedo (reflectivity) of Earth's surface and may lead to greater local and regional cooling. Similarly, snow accumulation also increases the albedo of land surfaces and often amplifies winter's effects.

Interannual Variation

Interannual climate variations, including droughts, floods, and other events, are caused by a complex array of factors and Earth system interactions. One important feature that plays a role in these variations is the periodic change of atmospheric and oceanic circulation patterns in the tropical Pacific region, collectively known as El Niño–Southern Oscillation (ENSO) variation. Although its primary climatic effects are concentrated in the tropical Pacific, ENSO has cascading effects that often extend to the Atlantic Ocean region, the interior of Europe and Asia, and the polar regions. These effects, called teleconnections, occur because alterations in low-latitude atmospheric circulation patterns in the Pacific region influence atmospheric circulation in adjacent and downstream systems. As a result, storm tracks are diverted and atmospheric pressure ridges (areas of high pressure) and troughs (areas of low pressure) are displaced from their usual patterns.

As an example, El Niño events occur when the easterly trade winds in the tropical Pacific weaken or reverse direction. This shuts down the upwelling of deep, cold waters off the west coast of South America, warms the eastern Pacific, and reverses the atmospheric pressure gradient in the western Pacific. As a result, air at the surface moves eastward from Australia and Indonesia toward the central Pacific and the Americas. These changes produce high rainfall and flash floods along the normally arid coast of Peru and severe drought in the normally wet regions of northern Australia and Indonesia. Particularly severe El Niño events lead to monsoon failure in the Indian Ocean region, resulting in intense drought in India and East Africa. At the same time, the westerlies and storm tracks are displaced toward the Equator, providing California and the desert Southwest of the United States with wet, stormy winter weather and causing winter conditions in the Pacific Northwest, which are typically wet, to become warmer and drier. Displacement of the westerlies also results in drought in northern China and from northeastern Brazil through sections of Venezuela. Long-term records of ENSO variation from historical documents, tree rings, and reef corals indicate that El Niño events occur, on average, every two to seven years. However, the frequency and intensity of these events vary through time.

The North Atlantic Oscillation (NAO) is another example of an interannual oscillation that produces important climatic effects within the Earth system and can influence climate throughout the Northern Hemisphere. This phenomenon results from variation in the pressure gradient, or the difference in atmospheric pressure between the subtropical high, usually situated between the Azores and Gibraltar, and the Icelandic low, centred between Iceland and Greenland. When the pressure gradient is steep due to a strong subtropical high and a deep Icelandic low (positive phase), northern Europe and northern Asia experience warm, wet winters with frequent strong winter storms. At the same time, southern Europe is dry. The eastern United States also experiences warmer, less snowy winters during positive NAO phases, although the effect is not as great as in Europe. The pressure gradient is dampened when NAO is in a negative mode—that is, when a weaker pressure gradient exists from the presence of a weak subtropical high and Icelandic low. When this happens, the Mediterranean region receives abundant winter rainfall, while northern Europe is cold and dry. The eastern United States is typically colder and snowier during a negative NAO phase.

During years when the North Atlantic Oscillation (NAO) is in its positive phase, the eastern United States, southeastern Canada, and northwestern Europe experience warmer winter temperatures, whereas colder temperatures are found in these locations during its negative phase. When the El

Niño/Southern Oscillation (ENSO) and NAO are both in their positive phase, European winters tend to be wetter and less severe; however, beyond this general tendency, the influence of the ENSO upon the NAO is not well understood.

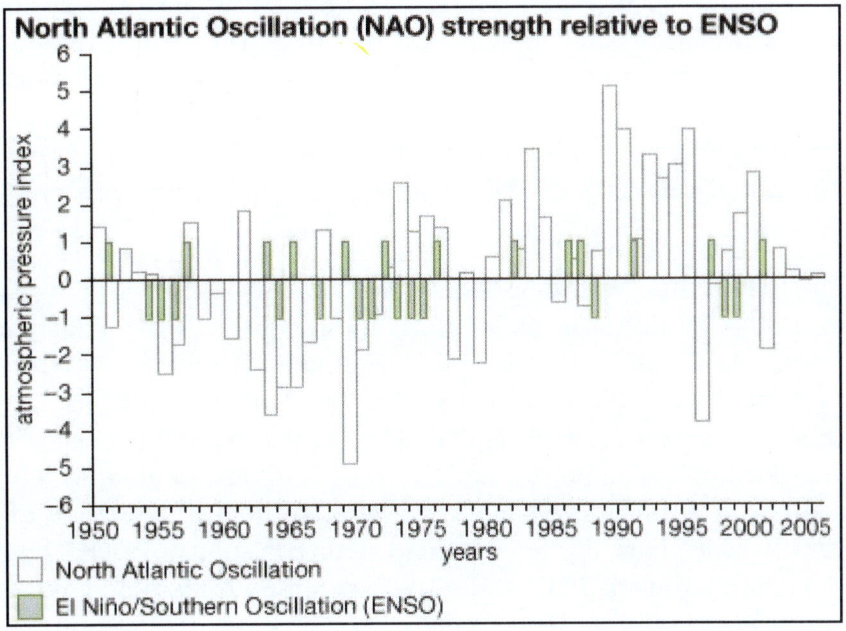

The ENSO and NAO cycles are driven by feedbacks and interactions between the oceans and atmosphere. Interannual climate variation is driven by these and other cycles, interactions among cycles, and perturbations in the Earth system, such as those resulting from large injections of aerosols from volcanic eruptions. One example of a perturbation due to volcanism is the 1991 eruption of Mount Pinatubo in the Philippines, which led to a decrease in the average global temperature of approximately 0.5 °C (0.9 °F) the following summer.

Decadal Variation

Climate varies on decadal timescales, with multiyear clusters of wet, dry, cool, or warm conditions. These multiyear clusters can have dramatic effects on human activities and welfare. For instance, a severe three-year drought in the late 16th century probably contributed to the destruction of Sir Walter Raleigh's "Lost Colony" at Roanoke Island in what is now North Carolina, and a subsequent seven-year drought led to high mortality at the Jamestown Colony in Virginia. Also, some scholars have implicated persistent and severe droughts as the main reason for the collapse of the Maya civilization in Mesoamerica between AD 750 and 950; however, discoveries in the early 21st century suggest that war-related trade disruptions played a role, possibly interacting with famines and other drought-related stresses.

Although decadal-scale climate variation is well documented, the causes are not entirely clear. Much decadal variation in climate is related to interannual variations. For example, the frequency and magnitude of ENSO change through time. The early 1990s were characterized by repeated El Niño events, and several such clusters have been identified as having taken place during the 20th century. The steepness of the NAO gradient also changes at decadal timescales; it has been particularly steep since the 1970s.

Recent research has revealed that decadal-scale variations in climate result from interactions between the ocean and the atmosphere. One such variation is the Pacific Decadal Oscillation (PDO), also referred to as the Pacific Decadal Variability (PDV), which involves changing sea surface temperatures (SSTs) in the North Pacific Ocean. The SSTs influence the strength and position of the Aleutian Low, which in turn strongly affects precipitation patterns along the Pacific Coast of North America. PDO variation consists of an alternation between "cool-phase" periods, when coastal Alaska is relatively dry and the Pacific Northwest relatively wet and "warm-phase" periods, characterized by relatively high precipitation in coastal Alaska and low precipitation in the Pacific Northwest. Tree ring and coral records, which span at least the last four centuries, document PDO variation.

A similar oscillation, the Atlantic Multidecadal Oscillation (AMO), occurs in the North Atlantic and strongly influences precipitation patterns in eastern and central North America. A warm-phase AMO (relatively warm North Atlantic SSTs) is associated with relatively high rainfall in Florida and low rainfall in much of the Ohio Valley. However, the AMO interacts with the PDO, and both interact with interannual variations, such as ENSO and NAO, in complex ways . Such interactions may lead to the amplification of droughts, floods, or other climatic anomalies. For example, severe droughts over much of the conterminous United States in the first few years of the 21st century were associated with warm-phase AMO combined with cool-phase PDO. The mechanisms underlying decadal variations, such as PDO and AMO, are poorly understood, but they are probably related to ocean-atmosphere interactions with larger time constants than interannual variations. Decadal climatic variations are the subject of intense study by climatologists and paleoclimatologists.

Climate Change since the Emergence of Civilization

Human societies have experienced climate change since the development of agriculture some 10,000 years ago. These climate changes have often had profound effects on human cultures and societies. They include annual and decadal climate fluctuations such as those described above, as well as large-magnitude changes that occur over centennial to multimillennial timescales. Such changes are believed to have influenced and even stimulated the initial cultivation and domestication of crop plants, as well as the domestication and pastoralization of animals. Human societies have changed adaptively in response to climate variations, although evidence abounds that certain societies and civilizations have collapsed in the face of rapid and severe climatic changes.

Centennial-scale Variation

Historical records as well as proxy records (particularly tree rings, corals, and ice cores) indicate that climate has changed during the past 1,000 years at centennial timescales; that is, no two centuries have been exactly alike. During the past 150 years, the Earth system has emerged from a period called the Little Ice Age, which was characterized in the North Atlantic region and elsewhere by relatively cool temperatures. The 20th century in particular saw a substantial pattern of warming in many regions. Some of this warming may be attributable to the transition from the Little Ice Age or other natural causes. However, many climate scientists believe that much of the 20th-century warming, especially in the later decades, resulted from atmospheric accumulation of greenhouse gases (especially carbon dioxide, CO_2).

The Little Ice Age is best known in Europe and the North Atlantic region, which experienced relatively cool conditions between the early 14th and mid-19th centuries. This was not a period of uniformly cool climate, since interannual and decadal variability brought many warm years. Furthermore, the coldest periods did not always coincide among regions; some regions experienced relatively warm conditions at the same time others were subjected to severely cold conditions. Alpine glaciers advanced far below their previous (and present) limits, obliterating farms, churches, and villages in Switzerland, France, and elsewhere. Frequent cold winters and cool, wet summers ruined wine harvests and led to crop failures and famines over much of northern and central Europe. The North Atlantic cod fisheries declined as ocean temperatures fell in the 17th century. The Norse colonies on the coast of Greenland were cut off from the rest of Norse civilization during the early 15th century as pack ice and storminess increased in the North Atlantic. The western colony of Greenland collapsed through starvation, and the eastern colony was abandoned. In addition, Iceland became increasingly isolated from Scandinavia.

The Little Ice Age was preceded by a period of relatively mild conditions in northern and central Europe. This interval, known as the Medieval Warm Period, occurred from approximately AD 1000 to the first half of the 13th century. Mild summers and winters led to good harvests in much of Europe. Wheat cultivation and vineyards flourished at far higher latitudes and elevations than today. Norse colonies in Iceland and Greenland prospered, and Norse parties fished, hunted, and explored the coast of Labrador and Newfoundland. The Medieval Warm Period is well documented in much of the North Atlantic region, including ice cores from Greenland. Like the Little Ice Age, this time was neither a climatically uniform period nor a period of uniformly warm temperatures everywhere in the world. Other regions of the globe lack evidence for high temperatures during this period.

Much scientific attention continues to be devoted to a series of severe droughts that occurred between the 11th and 14th centuries. These droughts, each spanning several decades, are well documented in tree-ring records across western North America and in the peatland records of the Great Lakes region. The records appear to be related to ocean temperature anomalies in the Pacific and Atlantic basins, but they are still inadequately understood. The information suggests that much of the United States is susceptible to persistent droughts that would be devastating for water resources and agriculture.

Millennial and Multimillennial Variation

The climatic changes of the past thousand years are superimposed upon variations and trends at both millennial timescales and greater. Numerous indicators from eastern North America and Europe show trends of increased cooling and increased effective moisture during the past 3,000 years. For example, in the Great Lakes–St. Lawrence regions along the U.S.-Canadian border, water levels of the lakes rose, peatlands developed and expanded, moisture-loving trees such as beech and hemlock expanded their ranges westward, and populations of boreal trees, such as spruce and tamarack, increased and expanded southward. These patterns all indicate a trend of increased effective moisture, which may indicate increased precipitation, decreased evapotranspiration due to cooling, or both. The patterns do not necessarily indicate a monolithic cooling event; more complex climatic changes probably occurred. For example, beech expanded northward and spruce southward during the past 3,000 years in both eastern North America and western Europe. The beech expansions may indicate milder winters or longer growing seasons, whereas the spruce expansions appear related to cooler, moister summers. Paleoclimatologists are applying a variety of approaches and proxies to help identify such changes in seasonal temperature and moisture during the Holocene Epoch.

Just as the Little Ice Age was not associated with cool conditions everywhere, so the cooling and moistening trend of the past 3,000 years was not universal. Some regions became warmer and drier during the same time period. For example, northern Mexico and the Yucatan experienced decreasing moisture in the past 3,000 years. Heterogeneity of this type is characteristic of climatic change, which involves changing patterns of atmospheric circulation. As circulation patterns change, the transport of heat and moisture in the atmosphere also changes. This fact explains the apparent paradox of opposing temperature and moisture trends in different regions.

The trends of the past 3,000 years are just the latest in a series of climatic changes that occurred over the past 11,700 years or so—the interglacial period referred to as the Holocene Epoch. At the start of the Holocene, remnants of continental glaciers from the last glaciation still covered much of eastern and central Canada and parts of Scandinavia. These ice sheets largely disappeared by 6,000 years ago. Their absence— along with increasing sea surface temperatures, rising sea levels (as glacial meltwater flowed into the world's oceans), and especially changes in the radiation budget of Earth's surface owing to Milankovitch variations (changes in the seasons resulting from periodic adjustments of Earth's orbit around the Sun)—affected atmospheric circulation. The diverse changes of the past 10,000 years across the globe are difficult to summarize in capsule, but some general highlights and large-scale patterns are worthy of note. These include the presence of early to mid-Holocene thermal maxima in various locations, variation in ENSO patterns, and an early to mid-Holocene amplification of the Indian Ocean monsoon.

Thermal Maxima

Many parts of the globe experienced higher temperatures than today some time during the early to mid-Holocene. In some cases the increased temperatures were accompanied by decreased moisture availability. Although the thermal maximum has been referred to in North America and elsewhere as a single widespread event (variously referred to as the "Altithermal", "Xerothermic Interval", "Climatic Optimum", or "Thermal Optimum"), it is now recognized that the periods of maximum temperatures varied among regions. For example, northwestern Canada experienced its highest temperatures several thousand years earlier than central or eastern North America. Similar heterogeneity is seen in moisture records. For instance, the record of the prairie-forest boundary in the Midwestern region of the United States shows eastward expansion of prairie in Iowa and Illinois 6,000 years ago (indicating increasingly dry conditions), whereas Minnesota's forests expanded westward into prairie regions at the same time (indicating increasing moisture). The Atacama Desert, located primarily in present-day Chile and Bolivia, on the western side of South America, is one of the driest places on Earth today, but it was much wetter during the early Holocene when many other regions were at their driest.

The primary driver of changes in temperature and moisture during the Holocene was orbital variation, which slowly changed the latitudinal and seasonal distribution of solar radiation on Earth's surface and atmosphere. However, the heterogeneity of these changes was caused by changing patterns of atmospheric circulation and ocean currents.

ENSO Variation in the Holocene

Because of the global importance of ENSO variation today, Holocene variation in ENSO patterns and intensity is under serious study by paleoclimatologists. The record is still fragmentary, but evidence from fossil corals, tree rings, lake records, climate modeling, and other approaches is accumulating that suggests that (1) ENSO variation was relatively weak in the early Holocene, (2) ENSO has undergone centennial to millennial variations in strength during the past 11,700 years, and (3) ENSO patterns and strength similar to those currently in place developed within the past 5,000 years. This evidence is particularly clear when comparing ENSO variation over the past 3,000 years to today's patterns. The causes of long-term ENSO variation are still being explored, but changes in solar radiation owing to Milankovitch variations are strongly implicated by modeling studies.

Amplification of the Indian Ocean Monsoon

Much of Africa, the Middle East, and the Indian subcontinent are under the strong influence of an annual climatic cycle known as the Indian Ocean monsoon. The climate of this region is highly seasonal, alternating between clear skies with dry air (winter) and cloudy skies with abundant rainfall (summer). Monsoon intensity, like other aspects of climate, is subject to interannual, decadal, and centennial variations, at least some of which are related to ENSO and other cycles. Abundant evidence exists for large variations in monsoon intensity during the Holocene Epoch. Paleontological and paleoecological studies show that large portions of the region experienced much greater precipitation during the early Holocene (11,700–6,000 years ago) than today. Lake and wetland sediments dating to this period have been found under the sands of parts of the Sahara Desert. These sediments contain fossils of elephants, crocodiles, hippopotamuses, and giraffes, together with pollen evidence of forest and woodland vegetation. In arid and semiarid parts of Africa, Arabia, and India, large and deep freshwater lakes occurred in basins that are now dry or are occupied by shallow, saline lakes. Civilizations based on plant cultivation and grazing animals, such as the Harappan civilization of northwestern India and adjacent Pakistan, flourished in these regions, which have since become arid.

These and similar lines of evidence, together with paleontological and geochemical data from marine sediments and climate-modeling studies, indicate that the Indian Ocean monsoon was greatly amplified during the early Holocene, supplying abundant moisture far inland into the African and Asian continents. This amplification was driven by high solar radiation in summer, which was approximately 7 percent higher 11,700 years ago than today and resulted from orbital forcing (changes in Earth's eccentricity, precession, and axial tilt). High summer insolation resulted in warmer summer air temperatures and lower surface pressure over continental regions and, hence, increased inflow of moisture-laden air from the Indian Ocean to the continental interiors. Modeling studies indicate that the monsoonal flow was further amplified by feedbacks involving the atmosphere, vegetation, and soils. Increased moisture led to wetter soils and lusher vegetation, which in turn led to increased precipitation and greater penetration of moist air into continental interiors. Decreasing summer insolation during the past 4,000–6,000 years led to the weakening of the Indian Ocean monsoon.

Climate Change since the Advent of Humans

The history of humanity—from the initial appearance of genus *Homo* over 2,000,000 years ago to the advent and expansion of the modern human species (*Homo sapiens*) beginning some 150,000 years ago—is integrally linked to climate variation and change. *Homo sapiens* has experienced nearly two full glacial-interglacial cycles, but its global geographical expansion, massive population increase, cultural diversification, and worldwide ecological domination began only during the last glacial period and accelerated during the last glacial-interglacial transition. The first bipedal apes appeared in a time of climatic transition and variation, and *Homo erectus*, an extinct species possibly ancestral to modern humans, originated during the colder Pleistocene Epoch and survived both the transition period and multiple glacial-interglacial cycles. Thus, it can be said that climate variation has been the midwife of humanity and its various cultures and civilizations.

Recent Glacial and Interglacial Periods

The most Recent Glacial Phase

With glacial ice restricted to high latitudes and altitudes, Earth 125,000 years ago was in an interglacial period similar to the one occurring today. During the past 125,000 years, however, the Earth system went through an entire glacial-interglacial cycle, only the most recent of many taking place over the last million years. The most recent period of cooling and glaciation began approximately 120,000 years ago. Significant ice sheets developed and persisted over much of Canada and northern Eurasia.

After the initial development of glacial conditions, the Earth system alternated between two modes, one of cold temperatures and growing glaciers and the other of relatively warm temperatures (although much cooler than today) and retreating glaciers. These Dansgaard-Oeschger (DO) cycles, recorded in both ice cores and marine sediments, occurred approximately every 1,500 years. A lower-frequency cycle, called the Bond cycle, is superimposed on the pattern of DO cycles; Bond cycles occurred every 3,000–8,000 years. Each Bond cycle is characterized by unusually cold conditions that take place during the cold phase of a DO cycle, the subsequent Heinrich event (which is a brief dry and cold phase), and the rapid warming phase that follows each Heinrich event. During each Heinrich event, massive fleets of icebergs were released into the North Atlantic, carrying rocks picked up by the glaciers far out to sea. Heinrich events are marked in marine sediments by conspicuous layers of iceberg-transported rock fragments.

Many of the transitions in the DO and Bond cycles were rapid and abrupt, and they are being studied intensely by paleoclimatologists and Earth system scientists to understand the driving mechanisms of such dramatic climatic variations. These cycles now appear to result from interactions between the atmosphere, oceans, ice sheets, and continental rivers that influence thermohaline circulation (the pattern of ocean currents driven by differences in water density, salinity, and temperature, rather than wind). Thermohaline circulation, in turn, controls ocean heat transport, such as the Gulf Stream.

Last Glacial Maximum

During the past 25,000 years, the Earth system has undergone a series of dramatic transitions. The most recent glacial period peaked 21,500 years ago during the Last Glacial Maximum, or LGM. At that time, the northern third of North America was covered by the Laurentide Ice Sheet, which extended as far south as Des Moines, Iowa; Cincinnati, Ohio; and New York City. The Cordilleran Ice Sheet covered much of western Canada as well as northern Washington, Idaho, and Montana in the United States. In Europe the Scandinavian Ice Sheet sat atop the British Isles, Scandinavia, northeastern Europe, and north-central Siberia. Montane glaciers were extensive in other regions, even at low latitudes in Africa and South America. Global sea level was 125 metres (410 feet) below modern levels, because of the long-term net transfer of water from the oceans to the ice sheets. Temperatures near Earth's surface in unglaciated regions were about 5 °C (9 °F) cooler than today. Many Northern Hemisphere plant and animal species inhabited areas far south of their present ranges. For example, jack pine and white spruce trees grew in northwestern Georgia, 1,000 km (600 miles) south of their modern range limits in the Great Lakes region of North America.

Last Deglaciation

The continental ice sheets began to melt back about 20,000 years ago. Drilling and dating of submerged fossil coral reefs provide a clear record of increasing sea levels as the ice melted. The most rapid melting began 15,000 years ago. For example, the southern boundary of the Laurentide Ice Sheet in North America was north of the Great Lakes and St. Lawrence regions by 10,000 years ago, and it had completely disappeared by 6,000 years ago.

The warming trend was punctuated by transient cooling events, most notably the Younger Dryas climate interval of 12,800–11,600 years ago. The climatic regimes that developed during the deglaciation period in many areas, including much of North America, have no modern analog (i.e., no regions exist with comparable seasonal regimes of temperature and moisture). For example, in the interior of North America, climates were much more continental (that is, characterized by warm summers and cold winters) than they are today. Also, paleontological studies indicate assemblages of plant, insect, and vertebrate species that do not occur anywhere today. Spruce trees grew with temperate hardwoods (ash, hornbeam, oak, and elm) in the upper Mississippi River and Ohio River regions. In Alaska, birch and poplar grew in woodlands, and there were very few of the spruce trees that dominate the present-day Alaskan landscape. Boreal and temperate mammals, whose geographic ranges are widely separated today, coexisted in central North America and Russia during this period of deglaciation. These unparalleled climatic conditions probably resulted from the combination of a unique orbital pattern that increased summer insolation and reduced winter insolation in the Northern Hemisphere and the continued presence of Northern Hemisphere ice sheets, which themselves altered atmospheric circulation patterns.

Climate Change and the Emergence of Agriculture

The first known examples of animal domestication occurred in western Asia between 11,000 and 9,500 years ago when goats and sheep were first herded, whereas examples of plant domestication date to 9,000 years ago when wheat, lentils, rye, and barley were first cultivated. This phase of technological increase occurred during a time of climatic transition that followed the last glacial period. A number of scientists have suggested that, although climate change imposed stresses on hunter-gatherer-forager societies by causing rapid shifts in resources, it also provided opportunities as new plant and animal resources appeared.

Glacial and Interglacial Cycles of the Pleistocene

The glacial period that peaked 21,500 years ago was only the most recent of five glacial periods in the last 450,000 years. In fact, the Earth system has alternated between glacial and interglacial regimes for more than two million years, a period of time known as the Pleistocene. The duration and severity of the glacial periods increased during this period, with a particularly sharp change occurring between 900,000 and 600,000 years ago. Earth is currently within the most recent interglacial period, which started 11,700 years ago and is commonly known as the Holocene Epoch.

The continental glaciations of the Pleistocene left signatures on the landscape in the form of glacial deposits and landforms; however, the best knowledge of the magnitude and timing of the various glacial and interglacial periods comes from oxygen isotope records in ocean sediments. These records provide both a direct measure of sea level and an indirect measure of global ice volume. Water molecules composed of a lighter isotope of oxygen, ^{16}O, are evaporated more readily than molecules bearing a heavier isotope, ^{18}O. Glacial periods are characterized by high ^{18}O concentrations and represent a net transfer of water, especially with ^{16}O, from the oceans to the ice sheets. Oxygen isotope records indicate that interglacial periods have typically lasted 10,000–15,000 years, and maximum glacial periods were of similar length. Most of the past 500,000 years—approximately 80 percent—have been spent within various intermediate glacial states that were warmer than glacial maxima but cooler than interglacials. During these intermediate times, substantial glaciers occurred over much of Canada and probably covered Scandinavia as well. These intermediate states were not constant; they were characterized by continual, millennial-scale climate variation. There has been no average or typical state for global climate during Pleistocene and Holocene times; the Earth system has been in continual flux between interglacial and glacial patterns.

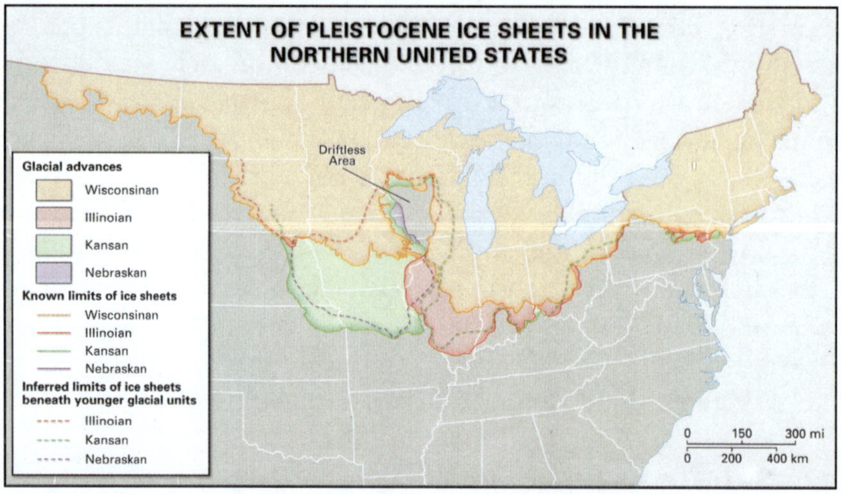

The blue areas are those that were covered by ice sheets in the past. The Kansan and Nebraskan sheets overlapped almost the same areas, and the Wisconsin and Illinoisan sheets covered approximately the same territory. In the high altitudes of the West are the Cordilleran ice sheets. An area at the junction of Wisconsin, Minnesota, Iowa, and Illinois was never entirely covered with ice.

The cycling of the Earth system between glacial and interglacial modes has been ultimately driven by orbital variations. However, orbital forcing is by itself insufficient to explain all of this variation, and Earth system scientists are focusing their attention on the interactions and feedbacks between the myriad components of the Earth system. For example, the initial development of a continental ice sheet increases albedo over a portion of Earth, reducing surface absorption of sunlight and leading to further cooling. Similarly, changes in terrestrial vegetation, such as the replacement of forests by tundra, feed back into the atmosphere via changes in both albedo and latent heat flux from evapotranspiration. Forests—particularly those of tropical and temperate areas, with their large leaf area—release great amounts of water vapour and latent heat through transpiration. Tundra plants, which are much smaller, possess tiny leaves designed to slow water loss; they release only a small fraction of the water vapour that forests do.

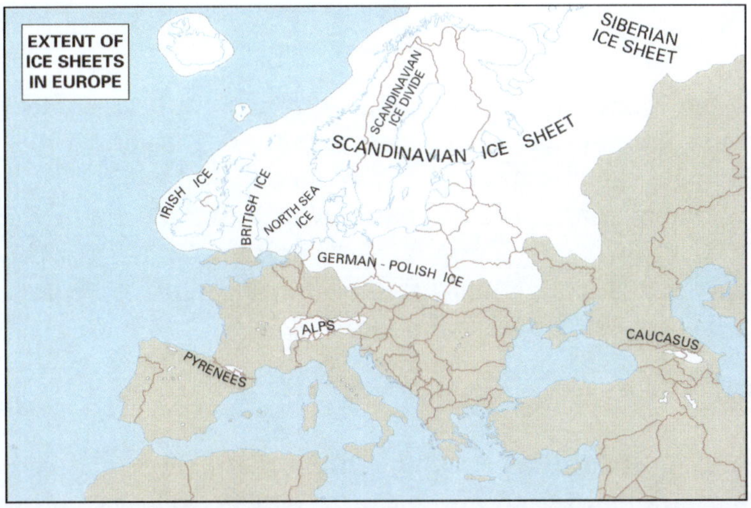

Europe, like North America, had four periods of glaciation. Successive ice caps reached limits that differed only slightly. The area covered by ice at any time is shown in white.

The discovery in ice core records that atmospheric concentrations of two potent greenhouse gases, carbon dioxide and methane, have decreased during past glacial periods and peaked during interglacials indicates important feedback processes in the Earth system. Reduction of greenhouse gas concentrations during the transition to a glacial phase would reinforce and amplify cooling already under way. The reverse is true for transition to interglacial periods. The glacial carbon sink remains a topic of considerable research activity. A full understanding of glacial-interglacial carbon dynamics requires knowledge of the complex interplay among ocean chemistry and circulation, ecology of marine and terrestrial organisms, ice sheet dynamics, and atmospheric chemistry and circulation.

Last Great Cooling

The Earth system has undergone a general cooling trend for the past 50 million years, culminating in the development of permanent ice sheets in the Northern Hemisphere about 2.75 million years ago. These ice sheets expanded and contracted in a regular rhythm, with each glacial maximum separated from adjacent ones by 41,000 years (based on the cycle of axial tilt). As the ice sheets waxed and waned, global climate drifted steadily toward cooler conditions characterized by increasingly severe glaciations and increasingly cool interglacial phases. Beginning around 900,000 years ago, the glacial-interglacial cycles shifted frequency. Ever since, the glacial peaks have been 100,000 years apart, and the Earth system has spent more time in cool phases than before. The 41,000-year periodicity has continued, with smaller fluctuations superimposed on the 100,000-year cycle. In addition, a smaller, 23,000-year cycle has occurred through both the 41,000-year and 100,000-year cycles.

The 23,000-year and 41,000-year cycles are driven ultimately by two components of Earth's orbital geometry: the equinoctial precession cycle (23,000 years) and the axial-tilt cycle (41,000 years).

Although the third parameter of Earth's orbit, eccentricity, varies on a 100,000-year cycle, its magnitude is insufficient to explain the 100,000-year cycles of glacial and interglacial periods of the past 900,000 years. The origin of the periodicity present in Earth's eccentricity is an important question in current paleoclimate research.

Climate Change through Geologic Time

The Earth system has undergone dramatic changes throughout its 4.5-billion-year history. These have included climatic changes diverse in mechanisms, magnitudes, rates, and consequences. Many of these past changes are obscure and controversial, and some have been discovered only recently. Nevertheless, the history of life has been strongly influenced by these changes, some of which radically altered the course of evolution. Life itself is implicated as a causative agent of some of these changes, as the processes of photosynthesis and respiration have largely shaped the chemistry of Earth's atmosphere, oceans, and sediments.

Cenozoic Climates

The Cenozoic Era—encompassing the past 65.5 million years, the time that has elapsed since the mass extinction event marking the end of the Cretaceous Period—has a broad range of climatic variation characterized by alternating intervals of global warming and cooling. Earth has experienced both extreme warmth and extreme cold during this period. These changes have been driven by tectonic forces, which have altered the positions and elevations of the continents as well as ocean passages and bathymetry. Feedbacks between different components of the Earth system (atmosphere, biosphere, lithosphere, cryosphere, and oceans in the hydrosphere) are being increasingly recognized as influences of global and regional climate. In particular, atmospheric concentrations of carbon dioxide have varied substantially during the Cenozoic for reasons that are poorly understood, though its fluctuation must have involved feedbacks between Earth's spheres.

Orbital forcing is also evident in the Cenozoic, although, when compared on such a vast era-level timescale, orbital variations can be seen as oscillations against a slowly changing backdrop of lower-frequency climatic trends. Descriptions of the orbital variations have evolved according to the growing understanding of tectonic and biogeochemical changes. A pattern emerging from recent paleoclimatologic studies suggests that the climatic effects of eccentricity, precession, and axial tilt have been amplified during cool phases of the Cenozoic, whereas they have been dampened during warm phases.

The meteor impact that occurred at or very close to the end of the Cretaceous came at a time of global warming, which continued into the early Cenozoic. Tropical and subtropical flora and fauna occurred at high latitudes until at least 40 million years ago, and geochemical records of marine sediments have indicated the presence of warm oceans. The interval of maximum temperature occurred during the late Paleocene and early Eocene epochs (58.7 million to 40.4 million years ago). The highest global temperatures of the Cenozoic occurred during the Paleocene-Eocene Thermal Maximum (PETM), a short interval lasting approximately 100,000 years. Although the underlying causes are unclear, the onset of the PETM about 56 million years ago was rapid, occurring within a few thousand years, and ecological consequences were large, with widespread extinctions in both marine and terrestrial ecosystems. Sea surface and continental air temperatures increased by more than 5 °C (9 °F) during the transition into the PETM. Sea surface temperatures in the high-latitude Arctic may have been as warm as 23 °C (73 °F), comparable to modern subtropical and warm-temperate seas. Following the PETM, global temperatures declined to pre-PETM levels, but they gradually increased to near-PETM levels over the next few million years during a period known as the Eocene Optimum. This temperature maximum was followed by a steady decline in global temperatures toward the Eocene-Oligocene boundary, which occurred about 33.9 million years ago. These changes are well-represented in marine sediments and in paleontological records from the continents, where vegetation zones moved Equator-ward. Mechanisms underlying the cooling trend are under study, but it is most likely that tectonic movements played an important role. This period saw the gradual opening of the sea passage between Tasmania and Antarctica, followed by the opening of the Drake Passage between South America and Antarctica. The latter, which isolated Antarctica within a cold polar sea, produced global effects on atmospheric and oceanic circulation. Recent evidence suggests that decreasing atmospheric concentrations of carbon dioxide during this period may have initiated a steady and irreversible cooling trend over the next few million years.

A continental ice sheet developed in Antarctica during the Oligocene Epoch, persisting until a rapid warming event took place 27 million years ago. The late Oligocene and early to mid-Miocene epochs (28.4 million to 13.8 million years ago) were relatively warm, though not nearly as warm as the Eocene. Cooling resumed 15 million years ago, and the Antarctic Ice Sheet expanded again to cover much of the continent. The cooling trend continued through the late Miocene and accelerated into the early Pliocene Epoch, 5.3 million years ago. During this period the Northern Hemisphere remained ice-free, and paleobotanical studies show cool-temperate Pliocene floras at high latitudes on Greenland and the Arctic Archipelago. The Northern Hemisphere glaciation, which began 3.2 million years ago, was driven by tectonic events, such as the closing of the Panama seaway and the uplift of the Andes, the Tibetan Plateau, and western parts of North America. These tectonic events led to changes in the circulation of the oceans and the atmosphere, which in turn fostered the development of persistent ice at high northern latitudes. Small-magnitude variations in carbon dioxide concentrations, which had been relatively low since at least the mid-Oligocene (28.4 million years ago), are also thought to have contributed to this glaciation.

Phanerozoic Climates

The Phanerozoic Eon (542 million years ago to the present), which includes the entire span of complex, multicellular life on Earth, has witnessed an extraordinary array of climatic states and transitions. The sheer antiquity of many of these regimes and events renders them difficult to understand in detail. However, a number of periods and transitions are well known, owing to good geological records and intense study by scientists. Furthermore, a coherent pattern of low-frequency climatic variation is emerging, in which the Earth system alternates between warm ("greenhouse") phases and cool ("icehouse") phases. The warm phases are characterized by high temperatures, high sea levels, and an absence of continental glaciers. Cool phases in turn are marked by low temperatures, low sea levels, and the presence of continental ice sheets, at least at high latitudes. Superimposed on these alternations are higher-frequency variations, where cool periods are embedded within greenhouse phases and warm periods are embedded within icehouse phases. For example, glaciers developed for a brief period (between 1 million and 10 million years) during the late Ordovician and early Silurian, in the middle of the early Paleozoic greenhouse phase (542 million to 350 million years ago). Similarly, warm periods with glacial retreat occurred within the late Cenozoic cool period during the late Oligocene and early Miocene epochs.

The Earth system has been in an icehouse phase for the past 30 million to 35 million years, ever since the development of ice sheets on Antarctica. The previous major icehouse phase occurred between about 350 million and 250 million years ago, during the Carboniferous and Permian periods of the late Paleozoic Era. Glacial sediments dating to this period have been identified in much of Africa as well as in the Arabian Peninsula, South America, Australia, India, and Antarctica. At the time, all these regions were part of Gondwana, a high-latitude supercontinent in the Southern Hemisphere. The glaciers atop Gondwana extended to at least 45° S latitude, similar to the latitude reached by Northern Hemisphere ice sheets during the Pleistocene. Some late Paleozoic glaciers extended even further Equator-ward—to 35° S. One of the most striking features of this time period are cyclothems, repeating sedimentary beds of alternating sandstone, shale, coal, and limestone. The great coal deposits of North America's Appalachian region, the American Midwest, and northern Europe are interbedded in these cyclothems, which may represent repeated transgressions (producing limestone) and retreats (producing shales and coals) of ocean shorelines in response to orbital variations.

The two most prominent warm phases in Earth history occurred during the Mesozoic and early Cenozoic eras (approximately 250 million to 35 million years ago) and the early and mid-Paleozoic (approximately 500 million to 350 million years ago). Climates of each of these greenhouse periods were distinct; continental positions and ocean bathymetry were very different, and terrestrial vegetation was absent from the continents until relatively late in the Paleozoic warm period. Both of these periods experienced substantial long-term climate variation and change; increasing evidence indicates brief glacial episodes during the mid-Mesozoic.

Understanding the mechanisms underlying icehouse-greenhouse dynamics is an important area of research, involving an interchange between geologic records and the modeling of the Earth system and its components. Two processes have been implicated as drivers of Phanerozoic climate change. First, tectonic forces caused changes in the positions and elevations of continents and the bathymetry of oceans and seas. Second, variations in greenhouse gases were also important drivers of climate, though at these long timescales they were largely controlled by tectonic processes, in which sinks and sources of greenhouse gases varied.

Climates of Early Earth

The pre-Phanerozoic interval, also known as Precambrian time, comprises some 88 percent of the time elapsed since the origin of Earth. The pre-Phanerozoic is a poorly understood phase of Earth system history. Much of the sedimentary record of the atmosphere, oceans, biota, and crust of the early Earth has been obliterated by erosion, metamorphosis, and subduction. However, a number of pre-Phanerozoic records have been found in various parts of the world, mainly from the later portions of the period. Pre-Phanerozoic Earth system history is an extremely active area of research, in part because of its importance in understanding the origin and early evolution of life on Earth. Furthermore, the chemical composition of Earth's atmosphere and oceans largely developed during this period, with living organisms playing an active role. Geologists, paleontologists, microbiologists, planetary geologists, atmospheric scientists, and geochemists are focusing intense efforts on understanding this period. Three areas of particular interest and debate are the "faint young Sun paradox", the role of organisms in shaping Earth's atmosphere, and the possibility that Earth went through one or more "snowball" phases of global glaciation.

Faint Young Sun Paradox

Astrophysical studies indicate that the luminosity of the Sun was much lower during Earth's early history than it has been in the Phanerozoic. In fact, radiative output was low enough to suggest that all surface water on Earth should have been frozen solid during its early history, but evidence shows that it was not. The solution to this "faint young Sun paradox" appears to lie in the presence of unusually high concentrations of greenhouse gases at the time, particularly methane and carbon dioxide. As solar luminosity gradually increased through time, concentrations of greenhouse gases would have to have been much higher than today. This circumstance would have caused Earth to heat up beyond life-sustaining levels. Therefore, greenhouse gas concentrations must have decreased proportionally with increasing solar radiation, implying a feedback mechanism to regulate greenhouse gases. One of these mechanisms might have been rock weathering, which is temperature-dependent and serves as an important sink for, rather than source of, carbon dioxide by removing sizable amounts of this gas from the atmosphere. Scientists are also looking to biological processes (many of which also serve as carbon dioxide sinks) as complementary or alternative regulating mechanisms of greenhouse gases on the young Earth.

Photosynthesis and Atmospheric Chemistry

The evolution by photosynthetic bacteria of a new photosynthetic pathway, substituting water (H_2O) for hydrogen sulfide (H_2S) as a reducing agent for carbon dioxide, had dramatic consequences for Earth system geochemistry. Molecular oxygen (O_2) is given off as a by-product of photosynthesis using the H_2O pathway, which is energetically more efficient than the more primitive H_2S pathway. Using H_2O as a reducing agent in this process led to the large-scale deposition of banded-iron formations, or BIFs, a source of 90 percent of present-day iron ores. Oxygen present in ancient oceans oxidized dissolved iron, which precipitated out of solution onto the ocean floors. This deposition process, in which oxygen was used up as fast as it was produced, continued for millions of years until most of the iron dissolved in the oceans was precipitated. By approximately 2 billion years ago, oxygen was able to accumulate in dissolved form in seawater and to outgas to the atmosphere. Although oxygen does not have greenhouse gas properties, it plays important indirect roles in Earth's climate, particularly in phases of the carbon cycle. Scientists are studying the role of oxygen and other contributions of early life to the development of the Earth system.

Snowball Earth Hypothesis

Geochemical and sedimentary evidence indicates that Earth experienced as many as four extreme cooling events between 750 million and 580 million years ago. Geologists have proposed that Earth's oceans and land surfaces were covered by ice from the poles to the Equator during these events. This "Snowball Earth" hypothesis is a subject of intense study and discussion. Two important questions arise from this hypothesis. First, how, once frozen, could Earth thaw? Second, how could life survive periods of global freezing? A proposed solution to the first question involves the outgassing of massive amounts of carbon dioxide by volcanoes, which could have warmed the planetary surface rapidly, especially given that major carbon dioxide sinks (rock weathering and photosynthesis) would have been dampened by a frozen Earth. A possible answer to the second question may lie in the existence of present-day life-forms within hot springs and deep-sea vents, which would have persisted long ago despite the frozen state of Earth's surface.

A counter-premise known as the "Slushball Earth" hypothesis contends that Earth was not completely frozen over. Rather, in addition to massive ice sheets covering the continents, parts of the planet (especially ocean areas near the Equator) could have been draped only by a thin, watery layer of ice amid areas of open sea. Under this scenario, photosynthetic organisms in low-ice or ice-free regions could continue to capture sunlight efficiently and survive these periods of extreme cold.

2

Causes of Climate Change

There are numerous elements which are responsible for affecting the climate, resulting in climate change. These include radiative forcing, deforestation, ozone depletion, global dimming, Milankovitch cycles, glacial period, orbital forcing, etc. This chapter has been carefully written to provide an in-depth understanding of the causes of climate change.

The earth's climate is dynamic and always changing through a natural cycle. What the world is more worried about is that the changes that are occurring today have been speeded up because of man's activities. These changes are being studied by scientists all over the world who are finding evidence from tree rings, pollen samples, ice cores, and sea sediments. The causes of climate change can be divided into two categories - those that are due to natural causes and those that are created by man.

Natural Causes

There are a number of natural factors responsible for climate change. Some of the more prominent ones are continental drift, volcanoes, ocean currents, the earth's tilt, and comets and meteorites.

Continental Drift

You may have noticed something peculiar about South America and Africa on a map of the world - don't they seem to fit into each other like pieces in a jigsaw puzzle? About 200 million years ago they were joined together! Scientists believe that back then, the earth was not as we see it today, but the continents were all part of one large landmass. Proof of this comes from the similarity between plant and animal fossils and broad belts of rocks found on the eastern coastline of South America and western coastline of Africa, which are now widely separated by the Atlantic Ocean. The discovery of fossils of tropical plants (in the form of coal deposits) in Antarctica has led to the conclusion that this frozen land at some time in the past, must have been situated closer to the equator, where the climate was tropical, with swamps and plenty of lush vegetation.

The continents that we are familiar with today were formed when the landmass began gradually drifting apart, millions of years back. This drift also had an impact on the climate because it changed the physical features of the landmass, their position and the position of water bodies. The separation of the landmasses changed the flow of ocean currents and winds, which affected the climate. This drift of the continents continues even today; the Himalayan range is rising by about 1 mm (millimeter) every year because the Indian land mass is moving towards the Asian land mass, slowly but steadily.

Volcanoes

When a volcano erupts it throws out large volumes of sulphur dioxide (SO_2), water vapour, dust, and ash into the atmosphere. Although the volcanic activity may last only a few days, yet the large volumes of gases and ash can influence climatic patterns for years. Millions of tonnes of sulphur dioxide gas can reach the upper levels of the atmosphere (called the stratosphere) from a major eruption. The gases and dust particles partially block the incoming rays of the sun, leading to cooling. Sulphur dioxide combines with water to form tiny droplets of sulphuric acid. These droplets are so small that many of them can stay aloft for several years. They are efficient reflectors of sunlight, and screen the ground from some of the energy that it would ordinarily receive from the sun. Winds in the upper levels of the atmopshere, called the stratosphere, carry the aerosols rapidly around the globe in either an easterly or westerly direction. Movement of aerosols north and south is always much slower. This should give you some idea of the ways by which cooling can be brought about for a few years after a major volcanic eruption.

Mount Pinatoba, in the Philippine islands erupted in April 1991 emitting thousands of tonnes of gases into the atmosphere. Volcanic eruptions of this magnitude can reduce the amount of solar radiation reaching the Earth's surface, lowering temperatures in the lower levels of the atmosphere (called the troposphere), and changing atmospheric circulation patterns. The extent to which this occurs is an ongoing debate.

Another striking example was in the year 1816, often referred to as "the year without a summer". Significant weather-related disruptions occurred in New England and in Western Europe with killing summer frosts in the United States and Canada. These strange phenomena were attributed to a major eruption of the Tambora volcano in Indonesia, in 1815.

Earth's Tilt

The earth makes one full orbit around the sun each year. It is tilted at an angle of 23.5° to the perpendicular plane of its orbital path. For one half of the year when it is summer, the northern hemisphere tilts towards the sun. In the other half when it is winter, the earth is tilted away from the sun. If there was no tilt we would not have experienced seasons. Changes in the tilt of the earth can affect the severity of the seasons - more tilt means warmer summers and colder winters; less tilt means cooler summers and milder winters.

The Earth's orbit is somewhat elliptical, which means that the distance between the earth and the Sun varies over the course of a year. We usually think of the earth's axis as being fixed, after all, it always seems to point toward Polaris (also known as the Pole Star and the North Star). Actually, it is not quite constant: the axis does move, at the rate of a little more than a half-degree each century. So Polaris has not always been, and will not always be, the star pointing to the North. When the pyramids were built, around 2500 BC, the pole was near the star Thuban (Alpha Draconis). This gradual change in the direction of the earth's axis, called precession is responsible for changes in the climate.

Ocean Currents

The oceans are a major component of the climate system. They cover about 71% of the Earth and absorb about twice as much of the sun's radiation as the atmosphere or the land surface. Ocean currents move vast amounts of heat across the planet - roughly the same amount as the atmosphere does. But the oceans are surrounded by land masses, so heat transport through the water is through channels.

Winds push horizontally against the sea surface and drive ocean current patterns. Certain parts of the world are influenced by ocean currents more than others. The coast of Peru and other adjoining regions are directly influenced by the Humboldt current that flows along the coastline of Peru. The El Niño event in the Pacific Ocean can affect climatic conditions all over the world.

Another region that is strongly influenced by ocean currents is the North Atlantic. If we compare places at the same latitude in Europe and North America the effect is immediately obvious. Take a closer look at this example - some parts of coastal Norway have an average temperature of -2°C in January and 14°C in July; while places at the same latitude on the Pacific coast of Alaska are far colder: -15°C in January and only 10°C in July. The warm current along the Norewgian coast keeps much of the Greenland-Norwegian Sea free of ice even in winter. The rest of the Arctic Ocean, even though it is much further south, remains frozen.

Ocean currents have been known to change direction or slow down. Much of the heat that escapes from the oceans is in the form of water vapour, the most abundant greenhouse gas on Earth. Yet, water vapor also contributes to the formation of clouds, which shade the surface and have a net cooling effect. Any or all of these phenomena can have an impact on the climate, as is believed to have happened at the end of the last Ice Age, about 14,000 years ago.

Human Causes

The Industrial Revolution in the 19th century saw the large-scale use of fossil fuels for industrial activities. These industries created jobs and over the years, people moved from rural areas to the cities. This trend is continuing even today. More and more land that was covered with vegetation has been cleared to make way for houses. Natural resources are being used extensively for construction, industries, transport, and consumption. Consumerism (our increasing want for material things) has increased by leaps and bounds, creating mountains of waste. Also, our population has increased to an incredible extent.

All this has contributed to a rise in greenhouse gases in the atmosphere. Fossil fuels such as oil, coal and natural gas supply most of the energy needed to run vehicles, generate electricity for industries, households, etc. The energy sector is responsible for about ¾ of the carbon dioxide emissions, 1/5 of the methane emissions and a large quantity of nitrous oxide. It also produces nitrogen oxides (NOx) and carbon monoxide (CO) which are not greenhouse gases but do have an influence on the chemical cycles in the atmosphere that produce or destroy greenhouse gases.

Greenhouse Gases and their Sources

Carbon dioxide is undoubtedly, the most important greenhouse gas in the atmosphere. Changes in land use pattern, deforestation, land clearing, agriculture, and other activities have all led to a rise in the emission of carbon dioxide.

Methane is another important greenhouse gas in the atmosphere. About ¼ of all methane emissions are said to come from domesticated animals such as dairy cows, goats, pigs, buffaloes, camels, horses, and sheep. These animals produce methane during the cud-chewing process. Methane is also released from rice or paddy fields that are flooded during the sowing and maturing periods. When soil is covered with water it becomes anaerobic or lacking in oxygen. Under such conditions, methane-producing bacteria and other organisms decompose organic matter in the soil to form methane. Nearly 90% of the paddy-growing area in the world is found in Asia, as rice is the staple food there. China and India, between them, have 80-90% of the world's rice-growing areas.

Methane is also emitted from landfills and other waste dumps. If the waste is put into an incinerator or burnt in the open, carbon dioxide is emitted. Methane is also emitted during the process of oil drilling, coal mining and also from leaking gas pipelines (due to accidents and poor maintenance of sites). A large amount of nitrous oxide emission has been attributed to fertilizer application. This in turn depends on the type of fertilizer that is used, how and when it is used and the methods of tilling that are followed. Contributions are also made by leguminous plants, such as beans and pulses that add nitrogen to the soil.

How we all Contribute Every Day

All of us in our daily lives contribute our bit to this change in the climate. Give these points a good, serious thought:

- Electricity is the main source of power in urban areas. All our gadgets run on electricity generated mainly from thermal power plants. These thermal power plants are run on fossil fuels (mostly coal) and are responsible for the emission of huge amounts of greenhouse gases and other pollutants.
- Cars, buses, and trucks are the principal ways by which goods and people are transported in most of our cities. These are run mainly on petrol or diesel, both fossil fuels.
- We generate large quantities of waste in the form of plastics that remain in the environment for many years and cause damage.
- We use a huge quantity of paper in our work at schools and in offices. Have we ever thought about the number of trees that we use in a day?
- Timber is used in large quantities for construction of houses, which means that large areas of forest have to be cut down.
- A growing population has meant more and more mouths to feed. Because the land area available for agriculture is limited (and in fact, is actually shrinking as a result of ecological degradation), high-yielding varieties of crop are being grown to increase the agricultural output from a given area of land. However, such high-yielding varieties of crops require large quantities of fertilizers; and more fertilizer means more emissions of nitrous oxide, both from the field into which it is put and the fertilizer industry that makes it. Pollution also results from the run-off of fertilizer into water bodies.

RADIATIVE FORCING

Radiative forcing or climate forcing is the difference between insolation (sunlight) absorbed by the Earth and energy radiated back to space. The influences that cause changes to the Earth's climate system altering Earth's radiative equilibrium, forcing temperatures to rise or fall, are called climate forcings. Positive radiative forcing means Earth receives more incoming energy from sunlight than it radiates to space. This net gain of energy will cause warming. Conversely, negative radiative forcing means that Earth loses more energy to space than it receives from the sun, which produces cooling.

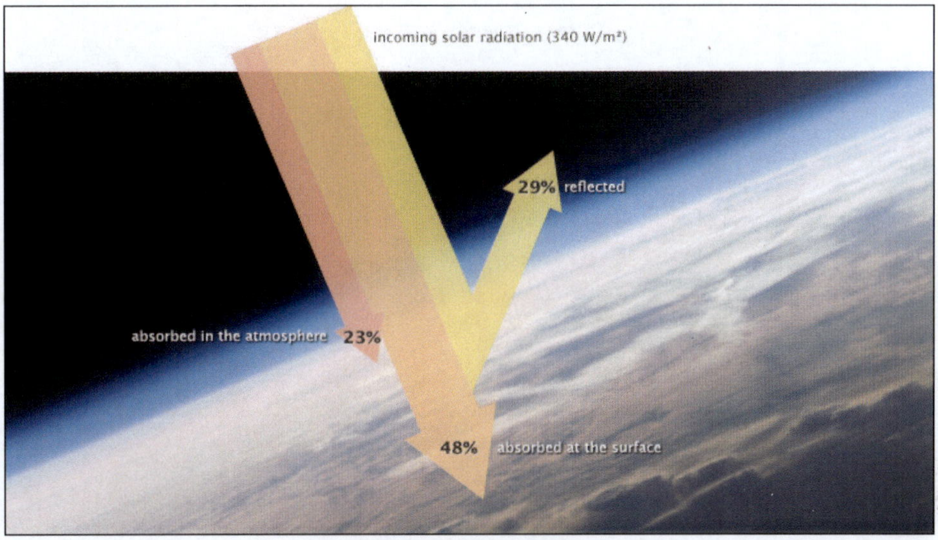

Incoming solar radiation.

Typically, radiative forcing is quantified at the tropopause or at the top of the atmosphere (often accounting for rapid adjustments in temperature) in units of watts per square meter of the Earth's surface. Positive forcing (incoming energy exceeding outgoing energy) warms the system, while negative forcing (outgoing energy exceeding incoming energy) cools it. Causes of radiative forcing include changes in insolation and the concentrations of radiatively active gases, commonly known as greenhouse gases, and aerosols.

Radiation Balance

Atmospheric gases only absorb some wavelengths of energy but are transparent to others. The absorption patterns of water vapor (blue peaks) and carbon dioxide (pink peaks) overlap in some wavelengths. Carbon dioxide is not as strong a greenhouse gas as water vapor, but it absorbs energy in wavelengths (12-15 micrometers) that water vapor does not, partially closing the "window" through which heat radiated by the surface would normally escape to space.

Almost all of the energy that affects Earth's climate is received as radiant energy from the Sun. The planet and its atmosphere absorb and reflect some of the energy, while long-wave energy is radiated back into space. The balance between absorbed and radiated energy determines the average global temperature. Because the atmosphere absorbs some of the re-radiated long-wave energy, the planet is warmer than it would be in the absence of the atmosphere.

The radiation balance is altered by such factors as the intensity of solar energy, reflectivity of clouds or gases, absorption by various greenhouse gases or surfaces and heat emission by various materials. Any such alteration is a radiative forcing, and changes the balance. This happens continuously as sunlight hits the surface, clouds and aerosols form, the concentrations of atmospheric gases vary and seasons alter the groundcover.

IPCC usage

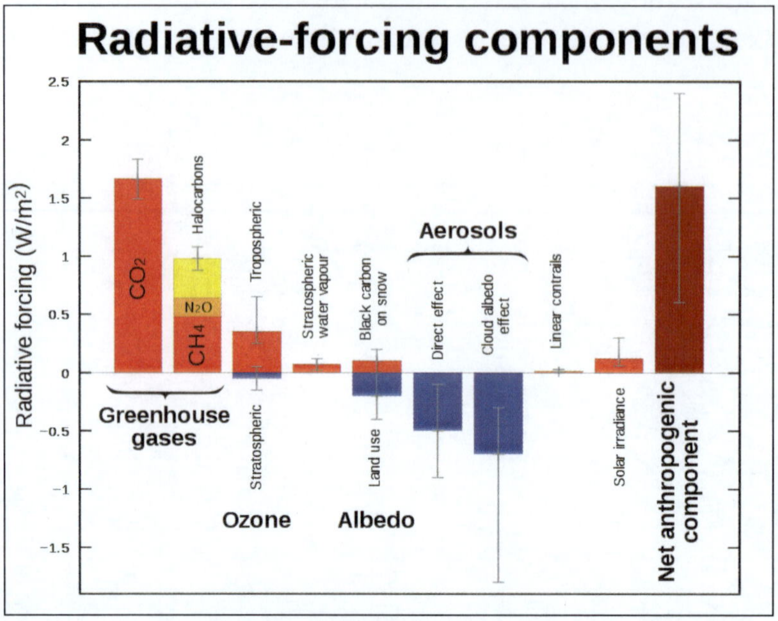

Radiative forcings, IPCC 2007.

The Intergovernmental Panel on Climate Change (IPCC) AR4 report defines radiative forcings as:

"Radiative forcing is a measure of the influence a factor has in altering the balance of incoming and outgoing energy in the Earth-atmosphere system and is an index of the importance of the factor as a potential climate change mechanism. In this report radiative forcing values are for changes relative to preindustrial conditions defined at 1750 and are expressed in Watts per square meter (W/m^2)".

In simple terms, radiative forcing is "the rate of energy change per unit area of the globe as measured at the top of the atmosphere". In the context of climate change, the term "forcing" is restricted to changes in the radiation balance of the surface-troposphere system imposed by external factors, with no changes in stratospheric dynamics, no surface and tropospheric feedbacks in operation (*i.e.*, no secondary effects induced because of changes in tropospheric motions or its thermodynamic state), and no dynamically induced changes in the amount and distribution of atmospheric water (vapour, liquid, and solid forms).

Sample Calculations

Solar Forcing

Radiative forcing (measured in watts per square meter) can be estimated in different ways for different components. For solar irradiance (i.e. "solar forcing"), the radiative forcing is simply the change in the average amount of solar energy absorbed per square meter of the Earth's area. Approximating the Earth as a sphere, the Earth's cross-sectional area exposed to the Sun (πr^2) is equal to 1/4 of the surface area of the Earth ($4\pi r^2$), the solar input per unit area is one quarter the change in solar intensity. Since some radiation is reflected, this must be multiplied by the fraction of incident sunlight that is absorbed, $F = (1 - R)$, where R is the reflectivity (albedo) of the Earth

—approximately 0.3, so *F* is approximately equal to 0.7. Thus, the solar forcing is the change in the solar intensity divided by 4 and multiplied by 0.7.

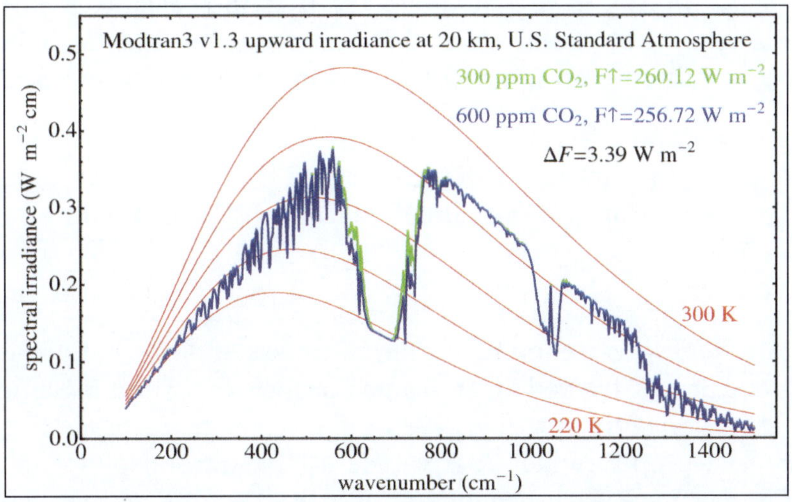

Radiative forcing for doubling CO_2, as calculated by radiative transfer code Modtran. Red lines are Planck curves.

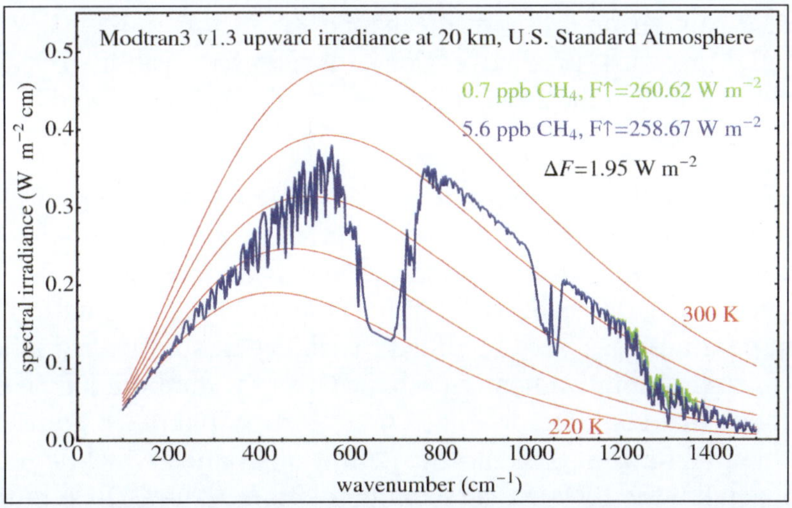

Radiative forcing for eight times increase of CH_4, as calculated by radiative transfer code Modtran.

Likewise, a change in albedo will produce a solar forcing equal to the change in albedo divided by 4 multiplied by the solar constant.

Forcing due to Atmospheric Gas

For a greenhouse gas, such as carbon dioxide, radiative transfer codes that examine each spectral line for atmospheric conditions can be used to calculate the change ΔF as a function of changing concentration. These calculations might be simplified into an algebraic formulation that is specific to that gas.

For instance, a proposed simplified first-order approximation expression for carbon dioxide would be:

$$\Delta F = 5.35 \times \ln \frac{C}{C_0} \text{ W m}^{-2}$$

where C is the CO_2 concentration in parts per million by volume and C_0 is the reference concentration. The relationship between carbon dioxide and radiative forcing is logarithmic, at concentrations up to around eight times the current value, and thus increased concentrations have a progressively smaller warming effect. Some claim that at higher concentrations, however, it becomes supra-logarithmic so that there is no saturation in the absorption of infrared radiation by CO_2.

A different formula might apply for other greenhouse gases such as methane and N2O (square-root dependence) or CFCs (linear), with coefficients that may be found *e.g.* in the IPCC reports. While recently a suggests a significant revision of methane IPCC formula.

Related Measures

Radiative forcing is a useful way to compare different causes of perturbations in a climate system. Other possible tools can be constructed for the same purpose: for example Shine say "recent experiments indicate that for changes in absorbing aerosols and ozone, the predictive ability of radiative forcing is much worse. we propose an alternative, the 'adjusted troposphere and stratosphere forcing'. We present GCM calculations showing that it is a significantly more reliable predictor of this GCM's surface temperature change than radiative forcing. It is a candidate to supplement radiative forcing as a metric for comparing different mechanisms". In this quote, GCM stands for "global circulation model", and the word "predictive" does not refer to the ability of GCMs to forecast climate change. Instead, it refers to the ability of the alternative tool proposed by the authors to help explain the system response.

Therefore, the concept of radiative forcing has been evolving from the initial proposal, named nowadays *instantaneous radiative forcing* (IRF), to other proposals that aims to relate better the radiative imbalance with global warming (global surface mean temperature). In this sense the adjusted radiative forcing, in its different calculation methodologies, estimates the imbalance once the stratosphere temperatures has been modified to achieve a radiative equilibrium in the stratosphere (in the sense of zero radiative heating rates). This new methodology is not estimating any *adjustment* or *feedback* that could be produced on the troposphere (in addition to stratospheric temperature adjustments), for that goal another definition, named *effective radiative forcing* has been introduced. In general the ERF is the recommendation of the CMIP6 radiative forcing analysis although the stratospherically adjusted methodologies are still being applied in those cases where the adjustments and feedbacks on the troposphere are considered not critical, like in the well mixed greenhouse gases and ozone. A methodology named *radiative kernel approach* allows to estimate the climate feedbacks within an offline calculation based on a linear approximation.

Climate Sensitivity

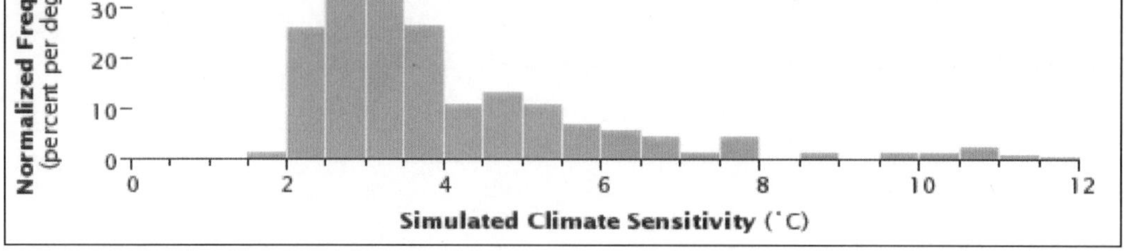

Frequency distribution of equilibrium climate sensitivity, based on model simulations. Each model simulation has a different guess at processes that scientist don't understand sufficiently well. Few of the simulations result in less than 2 °C of warming or significantly more than 4 °C. However the positive skew, which is also found in other studies, suggests that if carbon dioxide concentrations double, the probability of very large increases in temperature is greater than the probability of very small increases.

Climate sensitivity is the globally averaged temperature change in response to changes in radiative forcing, which can occur, for instance, due to increased levels of carbon dioxide (CO_2). Although the term climate sensitivity is usually used in the context of radiative forcing by CO_2, it is thought of as a general property of the climate system: the change in surface air temperature following a unit change in radiative forcing, and the climate sensitivity parameteris therefore expressed in units of °C/(W/m^2). The measure is approximately independent of the nature of the forcing (e.g. from greenhouse gases or solar variation). When climate sensitivity is expressed for a doubling of CO_2, its units are degrees Celsius (°C).

In the context of global warming, different measures of climate sensitivity are used. The *equilibrium climate sensitivity* (ECS) is the temperature increase that would result from sustained doubling of the concentration of carbon dioxide in Earth's atmosphere, after the Earth's energy budget and the climate system reach radiative equilibrium. The *transient climate response* (TCR) is the amount of temperature increase that might occur at the time when CO_2 doubles, having increased gradually by 1% each year. The *earth system sensitivity* (ESS) includes the effects of very-long-term Earth system feedback loops, such as changes in ice sheets or changes in the distribution of vegetative cover.

Climate sensitivity is typically estimated in three ways; by using observations taken during the industrial age, by using temperature and other data from the Earth's past and by modelling the climate system in computers. For coupled atmosphere-ocean global climate models the climate sensitivity is an emergent property; rather than being a model parameter it is a result of a combination of model physics and parameters. By contrast, simpler energy-balance models may have climate sensitivity as an explicit parameter.

Different Forms of Climate Sensitivity

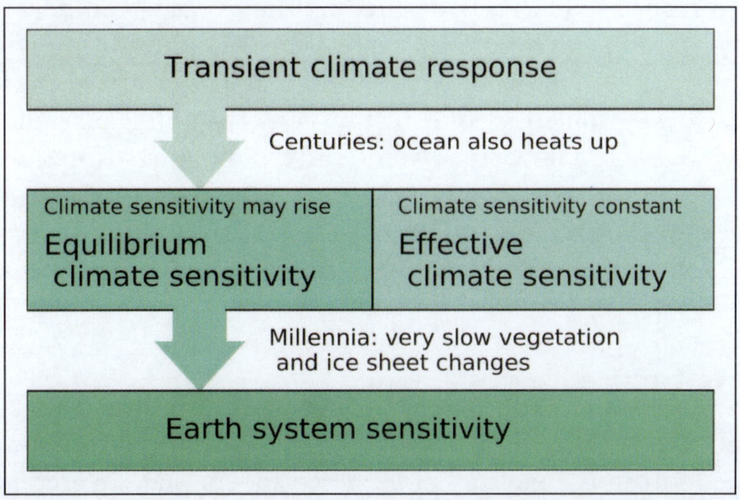

Schematic of how different measures of climate sensitivity relate to one another.

A component of climate sensitivity is directly due to radiative forcing, for instance by CO_2, and a further contribution arises from climate feedback, both positive and negative. Without feedbacks the radiative forcing of approximately 3.7 W/m^2, due to doubling CO_2 from the pre-industrial 280 ppm, would eventually result in roughly 1 °C global warming. This is easy to calculate and undisputed. The uncertainty is due entirely to feedbacks in the system: the water vapor feedback, the ice-albedo feedback, the cloud feedback, and the lapse rate feedback. Due to climate inertia, the climate sensitivity depends upon the timescale in which one is interested. The *transient response* is defined by scientists as the temperature response over human time scales of around 70 years, the *equilibrium climate sensitivity* over centuries, and finally the *Earth system sensitivity* after multiple millennia.

Equilibrium Climate Sensitivity

The equilibrium climate sensitivity (ECS) refers to the equilibrium change in global mean near-surface air temperature that would result from a sustained doubling of the atmospheric equivalent CO_2 concentration ($\Delta T_{2\times}$). A comprehensive model estimate of equilibrium sensitivity requires a very long model integration; fully equilibrating ocean temperatures requires the integration of thousands of model years, although it is possible to produce an estimate more quickly using the method of Gregory. As estimated by the IPCC Fifth Assessment Report (*AR5*), "there is high confidence that ECS is extremely unlikely less than 1°C and medium confidence that the ECS is likely between 1.5°C and 4.5°C and very unlikely greater than 6°C".

Effective Climate Sensitivity

The effective climate sensitivity is an estimate of equilibrium climate sensitivity using data from a climate system, either in a model or real-world observations, that is not yet in equilibrium. Estimation is done by using the assumption that the net effect of feedbacks as measured after a period of warming remains constant afterwards. This is not necessarily true, as feedbacks can change with time, or with the particular starting state or forcing history of the climate system.

Transient Climate Response

The transient climate response (TCR) is defined as the average temperature response over a twenty-year period centered at CO_2 doubling in a transient simulation with CO_2 increasing at 1% per year (compounded), i.e., 60 to 80 years following initiation of the increase in CO_2. The transient response is lower than the equilibrium sensitivity because the deep ocean, which takes many centuries to reach a new steady state after a perturbation, continues to serve as a sink for heat from the upper ocean. The IPCC literature assessment estimates that TCR likely lies between 1 °C and 2.5 °C. A related concept is the transient climate response to cumulative carbon emissions, which is the globally averaged surface temperature change per unit of CO_2 emitted.

Earth System Sensitivity

The Earth system sensitivity (ESS) includes the effects of slower feedback loops, such as the change in Earth's albedo from the melting of large ice sheets that covered much of the northern hemisphere during the last glacial maximum. These extra feedback loops make the ESS larger than the ECS – possibly twice as large. Data from Earth's history is used to estimate ESS, but climatic conditions were quite different which makes it difficult to infer information for future ESS. ESS includes the entire system except the carbon cycle. Changes in albedo as a result of vegetation changes are included.

Sensitivity to Nature of the Forcing

Radiative forcing from sources other than CO_2 can cause a higher or lower surface warming than a similar radiative forcing due to CO_2; the amount of feedback varies, mainly because these forcings are not uniformly distributed over the globe. Forcings that initially warm the northern hemisphere, land, or polar regions more strongly; are systematically more effective at changing temperatures than an equivalent amount of CO_2 whose forcing is more uniformly distributed over the globe. Several studies indicate that aerosols are more effective than CO_2 at changing global temperatures while volcanic forcing is less effective. Ignoring these factors causes lower estimates of climate sensitivity when using radiative forcing and temperature records from the historical period.

State Dependence

While climate sensitivity is defined as the sensitivity to any doubling of CO_2, there is evidence that the sensitivity of the climate system is not always constant. In the absence of sea ice, for instance, a positive ice-albedo feedback loop is lost, making the system less sensitive overall. Because of this, the climate system may warm by a different amount after a first doubling of CO_2 than after a second doubling. The effect of this is small or negligible in the first century after CO_2 is released into the atmosphere. Furthermore, climate sensitivity may change if tipping points are crossed. It is unlikely that climate sensitivity increases instantly; rather, it changes at the time scale of the subsystem that is undergoing the tipping point. The more sensitive a climate system is to increased greenhouse gases, the more likely it is to have decades when temperatures are much higher or much lower than the longer-term average.

Estimating Climate Sensitivity

Climate sensitivity is often evaluated in terms of the change in equilibrium temperature due to radiative forcing caused by the greenhouse effect. The radiative forcing, and hence the change in temperature, is proportional to the logarithm of the concentration of infrared-absorbing ("greenhouse") gases in the atmosphere, as quantified by Arrhenius in the 19th century. The sensitivity of temperature to atmospheric gasses, most notably CO_2, is often expressed in terms of the change in temperature per doubling of the concentration of the gas.

Svante Arrhenius was the first person to quantify global warming as a consequence of a doubling of CO_2. In his first paper on the matter, he estimated that global temperature would rise by around 5 to 6 °C (9.0 to 10.8 °F) if the quantity of CO_2 was doubled. In later work he revised this estimate to 4 °C (7.2 °F). Arrhenius used the observations of radiation emitted by the full moon made by the astronomer Samuel Pierpont Langley to estimate the amount of radiation that was absorbed by water vapour and CO_2. He then assumed relative humidity would stay the same under global warming as a representation of the water vapour feedback.

A committee on anthropogenic global warming convened in 1979 by the National Academy of Sciences and chaired by Jule Charney estimated climate sensitivity to be 3 °C (5.4 °F) with a tolerance of 1.5 °C (2.7 °F). Only two sets of models were available; one, due to Syukuro Manabe, showed a climate sensitivity of 2 °C (3.6 °F) and the other, due to James E. Hansen, showed a climate sensitivity of 4 °C (7.2 °F).

In 2008, climatologist Stefan Rahmstorf wrote, regarding the Charney report's original range of uncertainty; "At that time, this range was on very shaky ground. Since then, many vastly improved models have been developed by a number of climate research centers around the world. Current state-of-the-art climate models span a range of 2.6 to 4.1 °C (4.7 to 7.4 °F), most clustering around 3 °C (5.4 °F)".

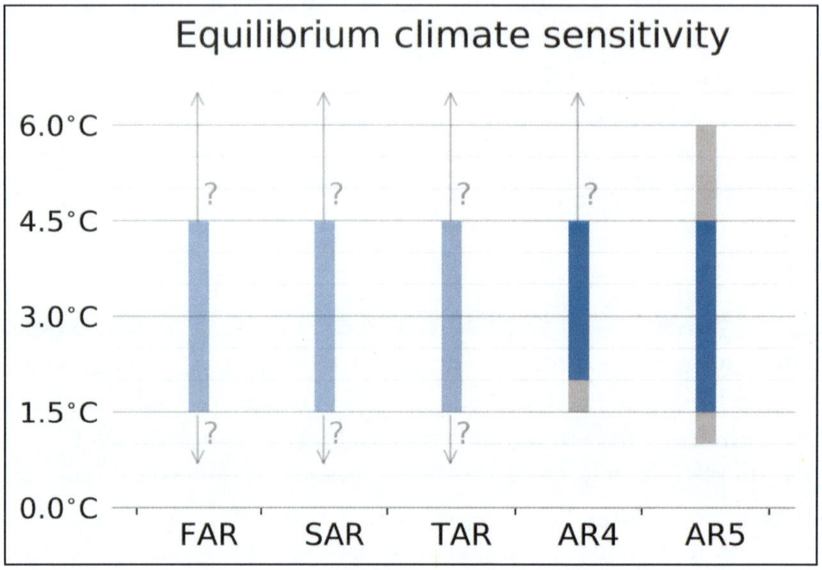

Historical estimates of climate sensitivity from the IPCC assessments. The first three reports gave a qualitative likely range, while the fourth and fifth assessment report formally quantified the uncertainty. The dark blue range is judged as being more than 66% likely.

After the Charney report, despite considerable progress in the understanding of the climate system, further assessments reported a similar range in climate sensitivity. The 1990 IPCC First Assessment Report estimated that equilibrium climate sensitivity to a doubling of CO_2 lay between 1.5 and 4.5 °C (2.7 and 8.1 °F), with a "best guess in the light of current knowledge" of 2.5 °C (4.5 °F). This report used models that had simplified representations of ocean dynamics. The IPCC supplementary report, 1992, which used full-ocean GCMs, saw "no compelling reason to warrant changing" from this estimate; and the IPCC Second Assessment Report said, "No strong reasons have emerged to change" these estimates. In these reports, much of the uncertainty was attributed to cloud processes. The 2001 IPCC TAR also retained this likely range.

Authors of the IPCC Fourth Assessment Report stated that confidence in estimates of equilibrium climate sensitivity had increased substantially since the Third Annual Report. IPCC authors concluded ECS is very likely to be greater than 1.5 °C (2.7 °F) and likely to lie in the range 2 to 4.5 °C (4 to 8.1 °F), with a most likely value of about 3 °C (5 °F). For fundamental physical reasons and data limitations, the IPCC stated a climate sensitivity higher than 4.5 °C (8.1 °F) cannot be ruled out but that agreement for these values with observations and "proxy" climate data is generally worse compared with values within the likely range.

The IPCC Fifth Assessment Report reverted to the earlier range of 1.5 to 4.5 °C (2.7 to 8.1 °F) (high confidence) because some estimates using industrial-age data came out low. They also stated that ECS is extremely unlikely to be less than 1 °C (1.8 °F) (high confidence), and is very unlikely to be greater than 6 °C (11 °F) (medium confidence). These values are estimated by combining the available data with expert judgement.

Using Industrial-age Data

Climate sensitivity can be estimated using observed temperature rise, observed ocean heat uptake, and modeled or observed radiative forcing. These data are linked though a simple energy-balance model to calculate climate sensitivity. Radiative forcing is often modeled, because satellites that measure it have not existed for the entire period. Estimates of climate sensitivity calculated from global energy constraints have consistently been lower than those calculated using other methods; estimates calculated using this method have been around 2 °C (3.6 °F) or lower.

Estimates of transient climate response (TCR) calculated from models and observational data can be reconciled if it is taken into account that fewer temperature measurements are taken in the polar regions, which warm more quickly than average. If only regions for which measurements are available are used in evaluating the model, differences in TCR estimates almost disappear.

Rahmstorf provides an informal example of the estimation of climate sensitivity using observations made since the pre-industrial era, from which the following is modified. Denote the sensitivity, i.e. the equilibrium increase in global mean temperature including the effects of feedbacks due to a sustained forcing by doubled CO_2 ($F_2 \times CO_2$; taken as 3.7 W/m^2), as S (°C). If Earth was to experience an equilibrium temperature change of ΔT (°C) due to a sustained forcing of ΔF (W/m^2), then:

$$S = \Delta T \times F_{2\times CO_2} / \Delta F.$$

The global temperature increase since the beginning of the industrial period is about 0.8 °C (1.4 °F), and the radiative forcing due to CO_2 and other long-lived greenhouse gases – mainly methane, nitrous oxide, and chlorofluorocarbons – emitted since that time is about 2.6 W/m^2. Neglecting other forcings and considering the temperature increase to be an equilibrium increase would lead to a sensitivity of about 1.1 °C (2.0 °F). However, ΔF also contains contributions from solar activity (+0.3 W/m^2), aerosols (−1.0 W/m^2), ozone (0.3 W/m^2), and other smaller influences, bringing the total forcing over the industrial period to 1.6 W/m^2 according to the best estimate of the IPCC AR4, with substantial uncertainty. The absence of equilibrium of the climate system must be accounted for by subtracting the planetary heat uptake rate H from the forcing; i.e.,

$$S = \Delta T \times F_{2\times CO_2} / (\Delta F - H).$$

Taking planetary heat uptake rate as the rate of ocean heat uptake estimated by the IPCC AR4 as 0.2 W/m^2, yields a value for S of 2.1 °C (3.8 °F).

Other Strategies

The industrial-age temperatures can also be used to determine a timescale of the climate system, which is theoretically linked to climate sensitivity. If the effective heat capacity of the climate system is known and the timescale estimated by using the autocorrelation of the measured temperature, an estimate of climate sensitivity can be derived. Determination of both the timescale and heat capacity has proven difficult.

Attempts to use the 11-year solar cycle to constrain the transient climate response have been made. Solar irradiance is about 0.9 W/m^2 brighter during solar maximum than during solar minimum, which correlated in measured average global temperature over the period 1959-2004. The solar minima in this period coincided with volcanic eruptions, which have a cooling effect on the global temperature. Because this causes a larger radiative forcing than the solar variations, it is questionable whether much information can be derived from the temperature variations.

Volcanic eruptions have also been used to find constraints on climate sensitivity. As the aerosols from a single volcanic eruption only last a couple of years in the atmosphere and the climate system's response to radiative forcing has inertia, it was found that only a lower bound on the transient climate sensitivity can be found.

Using Data from Earth's Past

Climate sensitivity can be estimated by using reconstructions of Earth's past temperatures and CO_2 levels. Different geological periods, for instance the warm Pliocene and the colder Pleistocene, are studied. Scientist seek periods that are in some sense analogous or informative to current climate change. As more information about the recent past becomes available, recent periods such as the Mid-holocene that occurred about 6,000 years ago and the Last Glacial Maximum (LGM) that took place about 21,000 years ago are often chosen.

The LGM was a period with a significantly lower global mean temperature than present day, and with a relatively well-known CO_2 concentration and radiative forcing in general. While orbital forcing was different from the present, this had little effect on mean annual temperatures. Different approaches to the task of estimating climate sensitivity from the LGM are taken. One approach is using estimates of global radiative forcing and temperature directly. The set of feedbacks active during the LGM, however, may be different than the feedbacks due to doubling CO_2, introducing additional uncertainty. In a different approach, a single model of intermediate complexity is run using a set of parameters so that each version has a different ECS. Those model versions that can best simulate the cooling during the LGM are thought to have the best ECS values. Others use an ensemble of different models.

Over the last 800,000 years, climate sensitivity has been found to be larger in cold periods than in warm periods. Climates further back in Earth's history are also used; an additional difficulty is that CO_2 concentrations cannot be readily obtained from ice cores so they must be estimated less directly. An estimate made using data from a major part of the Phanerozoic is consistent with sensitivities of current climate models and with other determinations. The Paleocene–Eocene Thermal Maximum provides a good opportunity to study the climate system when it is in a warm state.

Using Climate Models

Climate models of earth, for example the Coupled model intercomparison project (CMIP), are used to simulate the quantity of warming that will occur with rising CO_2 concentrations. The models are based on physical laws and represent the biosphere. Because of limited computer power, the physical laws have to be approximated, which leads to a wide range of estimates of climate sensitivity. Because the physical laws are programmed in a bottom-up way, climate sensitivity is an emergent property of these models.

In preparation for the 2021 6th IPCC report, a new generation of climate models are being developed: some show climate sensitivity around 5 °C (9.0 °F), meaning temperature can rise by 6.5 - 7 degree by 2100 in the worst socio-economic scenario ("SSP5 8.5 – rapid economic growth driven by fossil fuels without mitigation"). However the CMIP6 models have yet to be thoroughly independently analysed and researchers do not yet fully understand why some shows this higher sensitivity.

Constrained Models

Bottom-up modelling of the climate system can lead to a wide range of outcomes. Models are often run using different plausible parameters in their approximation of physical laws and the behaviour of the biosphere; a so-called perturbed physics ensemble. Alternatively, structurally different models developed at different institutions are put together, creating an ensemble. By selecting only those simulations that can simulate some part of the historical climate well, a constrained estimate of climate sensitivity can be made. One strategy is the placing of more trust in climate models that perform well in general.

Alternatively, specific metrics that are directly and physically linked to climate sensitivity are sought; examples of this are the global patterns of warming, the ability of the models to reproduce observed relative humidity in the tropics and sub-tropics, patterns of radiation, and the variability of temperature about long term historical warming. When using ensemble climate models developed in different institutions, many of these constrained estimates of ECS are slightly higher than 3 °C (5.4 °F); the models with ECS slightly above 3 °C (5.4 °F) perform better in these metrics than models with a low climate sensitivity.

DEFORESTATION

Deforestation is one of the main contributors to climate change. It comes in many forms: wildfire, agricultural clearcutting, livestock ranching, and logging for timber, among others. Forests cover 31% of the land area on Earth and annually 75,700 square kilometers (18.7 million acres) of forest is lost. Mass deforestation continues to threaten tropical forests, their biodiversity and the ecosystem services they provide. The main area of concern of deforestation is in tropical rainforests, since it is home to the majority of the biodiversity. Organisations such as World Wildlife Fund focus on the preservation of nature and the reduction of the most pressing threats to the diversity of life on Earth.

Deforestation is the second largest anthropogenic source of carbon dioxide to the atmosphere, after fossil fuel combustion. Deforestation and forest degradation contribute to atmospheric greenhouse gas emissions through combustion of forest biomass and decomposition of remaining plant material and soil carbon. It used to account for more than 20% of carbon dioxide emissions, but is currently around the 10% mark. By 2008, deforestation was 12% of total CO_2, or 15% if peatlands are included. These proportions are likely to have fallen since given the continued rise of fossil fuel use.

Averaged over all land and ocean surfaces, temperatures warmed roughly 1.53 °F (0.85 °C) between 1880 and 2012, according to the Intergovernmental Panel on Climate Change. In the Northern Hemisphere, 1983 to 2012 were the warmest 30-year period of the last 1400 years.

Causes of Deforestation

Lumber Industry

A large contributing factor to deforestation is the lumber industry. The reason behind this is due to logging since it is so common due to frequent wood production uses in many people's daily lives. A total of almost 4 million hectares (9.9×10^6 acres) of timber is harvested each year. In addition, the increasing demand for low costing timber products only supports to the lumber company to continue logging. The carbon emitted from the process of converting timber to wood products accounts for 15% of the carbon emissions in the environment. Deforestation is a main concern in tropical rainforest since it is home to millions of animals in the biodiversity. Not only does the lumber industry impact deforestation but also the environment due to deforestation drives on climate change.

Urbanisation

Urbanisation is the process of clearing a large area of land to build more living space. As a result of expanding land for residential and urban purposes, there is a significant amount of forest loss. As the human population continue to increase throughout the years, the demand and needs for necessities also increase. This means that more land is needed to be cleared in order to build more homes, recreational uses, and agriculture. In addition, this also promotes the industrialisation to grow and require more land to provide consumers with the food products demanded.

Livestock Ranching

Livestock ranching requires large portions of land to raise herds of animals and livestock crops for consumerism needs. Livestock ranching originated in Texas between 1820 and 1865 which were mainly driven by Mexican cowboys. Later, Texans drove the Mexicans out, while leaving the cattle behind. After the civil war, Texans begin rounding up the cattle and selling them around states such as California and New Orleans. According to Greenpeace, a non-governmental global environmental organisation, the cattle industry is responsible for a significant amount of methane emission. This is because speculators burn huge areas of rainforest for pasture. Rhett Butler says over 60% of land that is deforested become pasture for animals such as cattle and cows.

Agricultural Expansion

The number one largest cause of deforestation and acute degradation is agriculture. According to Wageningen University and Research Centre, more than 80% cause of deforestation is agriculture.

A continued increase of demand for timber and agricultural products are only critical indirect drives. Forest are giving way to plantation for coffee, tea, palm oil, rice, rubber, and many other highly demanded products. These rising demands for certain products and global trade arrangements causes forest conversions, which ultimately leads to soil erosion. The top soil oftentimes erodes after forests are being cleared which leads to sedimentation increase in rivers and streams. Over time, agricultural land degrades and becomes almost useless causing producers to find new productive lands.

Effect on Climate Change

Decrease in Biodiversity

A 2007 study conducted by the National Science Foundation found that biodiversity and genetic diversity are codependent—that diversity among species requires diversity within a species, and vice versa. "If any one type is removed from the system, the cycle can break down, and the community becomes dominated by a single species".

Decrease in Climate Services

Forests are nature's atmospheric carbon sink; plants take in atmospheric carbon dioxide (a greenhouse gas) and convert the carbon into sugars and plant materials through the process of photosynthesis. The carbon is stored within the trees, vegetation, and soil of the forests. Studies show that "intact forests," in fact, do sequester carbon. Examples of large forests that have a significant impact on the balance of carbon include the Amazonian and the Central African rainforests. However, deforestation disrupts the processes of carbon sequestration and affects localised climates. Additionally, cutting down trees plays a role in a positive feedback loop centred around climate change on a much larger scale, as studies are finding.

Burning or cutting down trees reverses the effects of carbon sequestration and releases greenhouse gases (including carbon dioxide) into the atmosphere. Furthermore, deforestation changes the landscape and reflectivity of earth's surface, i.e. decreasing albedo. This results in an increase in the absorption of light energy from the sun in the form of heat, enhancing global warming.

Implications on Soil and Water

Trees are a major source of carbon. It is estimated that the amount of carbon within the Amazon exceeds the ten year's worth of carbon released by human production. Unfortunately, since forests are often cleared by fire such as in slash and burn agriculture, the combustion process of wood release huge amounts of carbon dioxide into the atmosphere. The increase of atmospheric carbon is not the only consequence of deforestation, changes in soil properties could turn the soil itself into a carbon contributor. According to scientists at Yale University, clearing forests changes the environment of the microbial communities within the soil, and causes a loss of biodiversity in regards to the microbes since biodiversity is actually highly dependent on soil texture. Although the effect of deforestation has much more profound consequences on sandier soils compared to clay-like soils, the disruptions caused by deforestation ultimately reduces properties of soil such as hydraulic conductivity and water storage, thus reducing the efficiency of water and heat absorption. In a simulation of the deforestation process in the Amazon, researchers found that surface and soil temperatures increased by 1 to 3 degrees Celsius demonstrating the loss of the soil's ability to absorb radiation and moisture. Furthermore, soils that are rich in organic decay matter are more susceptible to fire, especially during long droughts. As a consequence of reduced evapotranspiration, precipitation is also reduced. This implies having a hotter and drier climate, and a longer dry season. This change in climate has drastic ecological and global impacts including increases in severity and frequency of fires, and disruption in the pollination process that will likely spread beyond the area of deforestation.

Counteracting Climate Change

Reforestation

Reforestation is the natural or intentional restocking of existing forests and woodlands that have been depleted, usually through deforestation. It is the reestablishment of forest cover either naturally or artificially. Similar to the other methods of forestation, reforestation can be very effective because a single tree can absorb as much as 22 kilograms (48 lb) of carbon dioxide per year and can sequester 0.91 tonnes (1 short ton) of carbon dioxide by the time it reaches 40 years old.

Afforestation

Afforestation is the planting of trees where there was no previous tree coverage. Degradation of forest ultimately leads to a decrease in oxygen and a sufficient increase of carbon dioxide. In order to make up for the loss, more trees are being planted. As a result, the amount of carbon dioxide in the atmosphere could significantly decrease. According to a scientific research, plantation forest could absorb more carbon dioxide than natural forest since they grow faster leading to a higher absorbance rate. The process is usually encouraged by governments because they want it to lead to a decrease in carbon dioxide and because it increases the aesthetics of the area. Although, it could lead to infringing upon ecosystems and create complications in environments that previously did not have tree coverage or forests.

China

Although China has set official goals for reforestation, these goals were set for an 80-year time horizon and were not significantly met by 2008. China is trying to correct these problems with projects such as the Green Wall of China, which aims to replant forests and halt the expansion of the Gobi Desert. A law promulgated in 1981 requires that every school student over the age of 11 plant at least one tree per year. But average success rates, especially in state-sponsored plantings, remains relatively low. And even the properly planted trees have had great difficulty surviving the combined impacts of prolonged droughts, pest infestation and fires. Nonetheless, China currently has the highest afforestation rate of any country or region in the world, with 4.77 million hectares (47,000 square kilometres) of afforestation in 2008.

Japan

The primary goal of afforestation projects in Japan is to develop the forest structure of the nation and to maintain the biodiversity found in the Japanese wilderness. The Japanese temperate rainforest is scattered throughout the Japanese archipelago and is home to many endemic species that are not naturally found anywhere else. As development of the country's caused a decline in forest cover, a reduction in biodiversity was seen in those areas.

Agroforestry

Agroforestry or agro-sylviculture is a land use management system in which combinations of trees or shrubs are grown around or among crops or pastureland. It combines agricultural and forestry technologies to create more diverse, productive, profitable, healthy, and sustainable land-use systems. There are many benefits to agroforestry such as increasing farm profitability. In addition, agroforestry helps to preserve and protect natural resources such as controlling soil erosions, creating habitat for the wildlife, and managing animal waste.

Reduce Emissions from Deforestation and Forest Degradation

Recognition of the negative impacts of deforestation and of the overwhelming evidence of global warming has led to the development of international policy surrounding the conservation of forests. One attempt towards fighting climate change globally is the Reducing Emissions for Deforestation and Forest Degradation (REDD+) efforts, and a few countries are already starting to implement and analyse ways to protect standing trees.

In the case of the Bac Kan province in Vietnam, researchers came up with systems to encourage leaving forests intact while also meeting international, national, and individual investments successfully. Their methods included "benefit-distribution systems" and dividends for ecosystem services. The researchers hope that their results "can be replicated and directly contribute to reducing carbon emissions globally".

OZONE DEPLETION

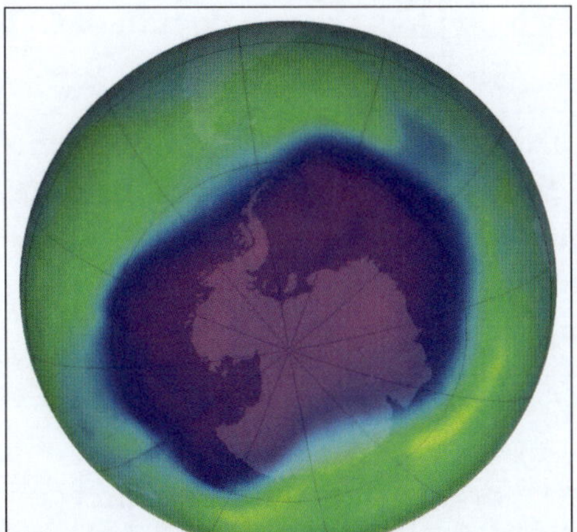

Image of the largest Antarctic ozone hole ever recorded over the South Pole in September 2006.

Ozone depletion consists of two related events observed since the late 1970s: a steady lowering of about four percent in the total amount of ozone in Earth's atmosphere (the ozone layer), and a much larger springtime decrease in stratospheric ozone around Earth's polar regions. The latter phenomenon is referred to as the ozone hole. There are also springtime polar tropospheric ozone depletion events in addition to these stratospheric events.

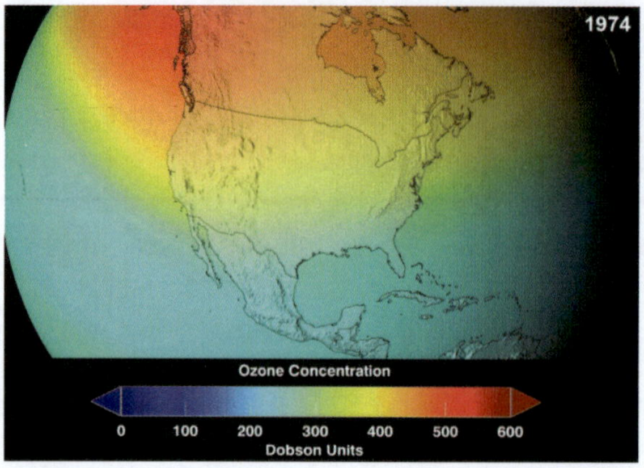

NASA projections of stratospheric ozone concentrations if chlorofluorocarbons had not been banned.

The main cause of ozone depletion and the ozone hole is manufactured chemicals, especially manufactured halocarbon refrigerants, solvents, propellants and foam-blowing agents (chlorofluorocarbons (CFCs), HCFCs, halons), referred to as ozone-depleting substances (ODS). These compounds are transported into the stratosphere by turbulent mixing after being emitted from the surface, mixing much faster than the molecules can settle. Once in the stratosphere, they release halogen atoms through photodissociation, which catalyze the breakdown of ozone (O_3) into oxygen (O_2). Both types of ozone depletion were observed to increase as emissions of halocarbons increased.

Ozone depletion and the ozone hole have generated worldwide concern over increased cancer risks and other negative effects. The ozone layer prevents most harmful UV wavelengths of ultraviolet light (UV light) from passing through the Earth's atmosphere. These wavelengths cause skin cancer, sunburn and cataracts, which were projected to increase dramatically as a result of thinning ozone, as well as harming plants and animals. These concerns led to the adoption of the Montreal Protocol in 1987, which bans the production of CFCs, halons and other ozone-depleting chemicals.

The ban came into effect in 1989. Ozone levels stabilized by the mid-1990s and began to recover in the 2000s. Recovery is projected to continue over the next century, and the ozone hole is expected to reach pre-1980 levels by around 2075. The Montreal Protocol is considered the most successful international environmental agreement to date.

Ozone Cycle

Three forms (or allotropes) of oxygen are involved in the ozone-oxygen cycle: oxygen atoms (O or atomic oxygen), oxygen gas (O_2 or diatomic oxygen), and ozone gas (O_3 or triatomic oxygen). Ozone is formed in the stratosphere when oxygen molecules photodissociate after absorbing ultraviolet photons. This converts a single O_2 into two atomic oxygen radicals. The atomic oxygen radicals then combine with separate O_2 molecules to create two O_3 molecules. These ozone molecules absorb ultraviolet (UV) light, following which ozone splits into a molecule of O_2 and an oxygen atom. The oxygen atom then joins up with an oxygen molecule to regenerate ozone. This is a continuing process that terminates when an oxygen atom recombines with an ozone molecule to make two O_2 molecules.

$$O + O_3 \rightarrow 2\,O_2$$

The total amount of ozone in the stratosphere is determined by a balance between photochemical production and recombination.

Ozone can be destroyed by a number of free radical catalysts; the most important are the hydroxyl radical (OH·), nitric oxide radical (NO·), chlorine radical (Cl·) and bromine radical (Br·). The dot is a notation to indicate that each species has an unpaired electron and is thus extremely reactive. All of these have both natural and man-made sources; at the present time, most of the OH· and NO· in the stratosphere is naturally occurring, but human activity has drastically increased the levels of chlorine and bromine. These elements are found in stable organic compounds, especially chlorofluorocarbons, which can travel to the stratosphere without being destroyed in the troposphere due to their low reactivity. Once in the stratosphere, the Cl and Br atoms are released from the parent compounds by the action of ultraviolet light, e.g.

$$CFCl_3 + \text{electromagnetic radiation} \rightarrow Cl\cdot + \cdot CFCl_2$$

Ozone is a highly reactive molecule that easily reduces to the more stable oxygen form with the assistance of a catalyst. Cl and Br atoms destroy ozone molecules through a variety of catalytic cycles. In the simplest example of such a cycle, a chlorine atom reacts with an ozone molecule (O_3), taking an oxygen atom to form chlorine monoxide (ClO) and leaving an oxygen molecule (O_2). The ClO can react with a second molecule of ozone, releasing the chlorine atom and yielding two molecules of oxygen. The chemical shorthand for these gas-phase reactions is:

- $Cl\cdot + O_3 \rightarrow ClO + O_2$ A chlorine atom removes an oxygen atom from an ozone molecule to make a ClO molecule.
- $ClO + O_3 \rightarrow Cl\cdot + 2\,O_2$ This ClO can also remove an oxygen atom from another ozone molecule; the chlorine is free to repeat this two-step cycle.

The overall effect is a decrease in the amount of ozone, though the rate of these processes can be decreased by the effects of null cycles. More complicated mechanisms have also been discovered that lead to ozone destruction in the lower stratosphere.

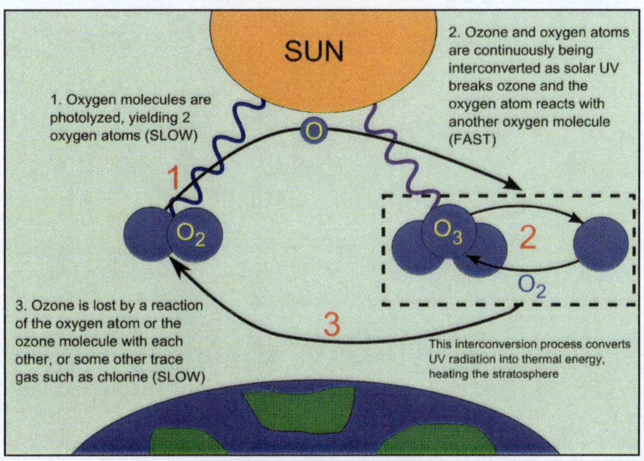

The ozone cycle.

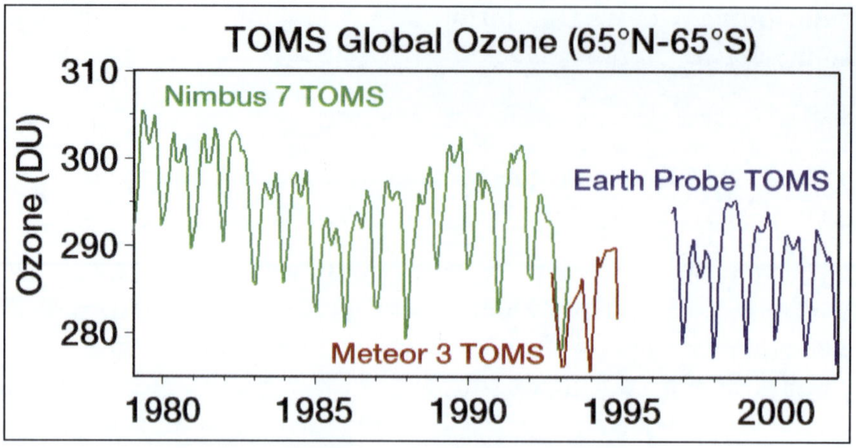

Global monthly average total ozone amount.

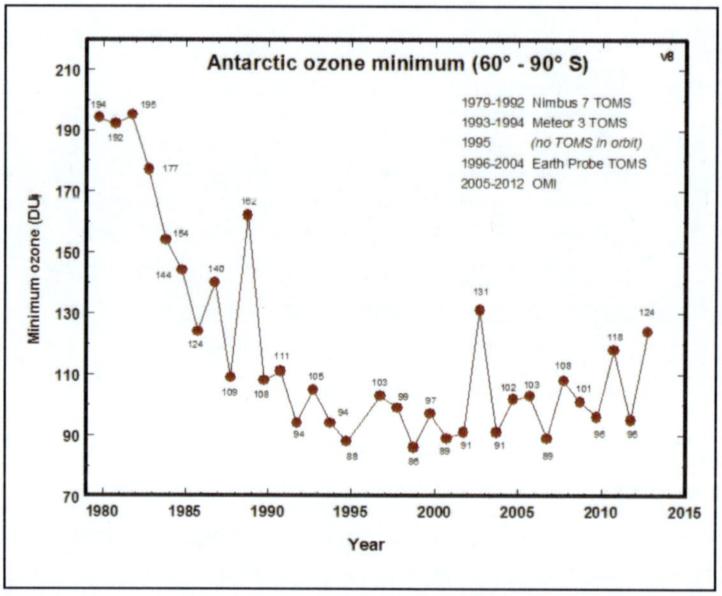

Lowest value of ozone measured by TOMS each year in the ozone hole.

A single chlorine atom would continuously destroy ozone (thus a catalyst) for up to two years (the time scale for transport back down to the troposphere) were it not for reactions that remove them from this cycle by forming reservoir species such as hydrogen chloride (HCl) and chlorine nitrate ($ClONO_2$). Bromine is even more efficient than chlorine at destroying ozone on a per atom basis, but there is much less bromine in the atmosphere at present. Both chlorine and bromine contribute significantly to overall ozone depletion. Laboratory studies have also shown that fluorine and iodine atoms participate in analogous catalytic cycles. However, fluorine atoms react rapidly with water and methane to form strongly bound HF in the Earth's stratosphere, while organic molecules containing iodine react so rapidly in the lower atmosphere that they do not reach the stratosphere in significant quantities.

A single chlorine atom is able to react with an average of 100,000 ozone molecules before it is removed from the catalytic cycle. This fact plus the amount of chlorine released into the atmosphere yearly by chlorofluorocarbons (CFCs) and hydrochlorofluorocarbons (HCFCs) demonstrates the danger of CFCs and HCFCs to the environment.

Observations on Ozone Layer Depletion

The ozone hole is usually measured by reduction in the total *column ozone* above a point on the Earth's surface. This is normally expressed in Dobson units; abbreviated as "DU". The most prominent decrease in ozone has been in the lower stratosphere. Marked decreases in column ozone in the Antarctic spring and early summer compared to the early 1970s and before have been observed using instruments such as the Total Ozone Mapping Spectrometer (TOMS).

Reductions of up to 70 percent in the ozone column observed in the austral (southern hemispheric) spring over Antarctica and first reported in 1985 are continuing. Antarctic total column ozone in September and October have continued to be 40–50 percent lower than pre-ozone-hole values since the 1990s. A gradual trend toward "healing" was reported in 2016. In 2017, NASA announced that the ozone hole was the weakest since 1988 because of warm stratospheric conditions. It is expected to recover around 2070.

The amount lost is more variable year-to-year in the Arctic than in the Antarctic. The greatest Arctic declines are in the winter and spring, reaching up to 30 percent when the stratosphere is coldest.

Reactions that take place on polar stratospheric clouds (PSCs) play an important role in enhancing ozone depletion. PSCs form more readily in the extreme cold of the Arctic and Antarctic stratosphere. This is why ozone holes first formed, and are deeper, over Antarctica. Early models failed to take PSCs into account and predicted a gradual global depletion, which is why the sudden Antarctic ozone hole was such a surprise to many scientists.

It is more accurate to speak of ozone depletion in middle latitudes rather than holes. Total column ozone declined below pre-1980 values between 1980 and 1996 for mid-latitudes. In the northern mid-latitudes, it then increased from the minimum value by about two percent from 1996 to 2009 as regulations took effect and the amount of chlorine in the stratosphere decreased. In the Southern Hemisphere's mid-latitudes, total ozone remained constant over that time period. There are no significant trends in the tropics, largely because halogen-containing compounds have not had time to break down and release chlorine and bromine atoms at tropical latitudes.

Large volcanic eruptions have been shown to have substantial albeit uneven ozone-depleting effects, as observed with the 1991 eruption of Mt. Pinotubo in the Philippines.

Ozone depletion also explains much of the observed reduction in stratospheric and upper tropospheric temperatures. The source of the warmth of the stratosphere is the absorption of UV radiation by ozone, hence reduced ozone leads to cooling. Some stratospheric cooling is also predicted from increases in greenhouse gases such as CO_2 and CFCs themselves; however, the ozone-induced cooling appears to be dominant.

Predictions of ozone levels remain difficult, but the precision of models' predictions of observed values and the agreement among different modeling techniques have increased steadily. The World Meteorological Organization Global Ozone Research and Monitoring Project—Report No. 44 comes out strongly in favor of the Montreal Protocol, but notes that a UNEP 1994 Assessment overestimated ozone loss for the 1994–1997 period.

Compounds in the Atmosphere

CFCs and Related Compounds in the Atmosphere

Chlorofluorocarbons (CFCs) and other halogenated ozone depleting substances (ODS) are mainly responsible for man-made chemical ozone depletion. The total amount of effective halogens (chlorine and bromine) in the stratosphere can be calculated and are known as the equivalent effective stratospheric chlorine (EESC).

CFCs were invented by Thomas Midgley, Jr. in the 1930s. They were used in air conditioning and cooling units, as aerosol spray propellants prior to the 1970s, and in the cleaning processes of delicate electronic equipment. They also occur as by-products of some chemical processes. No significant natural sources have ever been identified for these compounds—their presence in the atmosphere is due almost entirely to human manufacture. When such ozone-depleting chemicals reach the stratosphere, they are dissociated by ultraviolet light to release chlorine atoms. The chlorine atoms act as a catalyst, and each can break down tens of thousands of ozone molecules before being removed from the stratosphere. Given the longevity of CFC molecules, recovery times are measured in decades. It is calculated that a CFC molecule takes an average of about five to seven years to go from the ground level up to the upper atmosphere, and it can stay there for about a century, destroying up to one hundred thousand ozone molecules during that time.

1,1,1-Trichloro-2,2,2-trifluoroethane, also known as CFC-113a, is one of four man-made chemicals newly discovered in the atmosphere by a team at the University of East Anglia. CFC-113a is the only known CFC whose abundance in the atmosphere is still growing. Its source remains a mystery, but illegal manufacturing is suspected by some. CFC-113a seems to have been accumulating unabated since 1960. Between 2010 and 2012, emissions of the gas jumped by 45 percent.

Computer Modeling

Scientists have attributed ozone depletion to the increase of man-made (anthropogenic) halogen compounds from CFCs by combining observational data with computer models. These complex chemistry transport models (e.g. SLIMCAT, CLaMS—Chemical Lagrangian Model of the Stratosphere) work by combining measurements of chemicals and meteorological fields with chemical reaction rate constants. They identify key chemical reactions and transport processes that bring CFC photolysis products into contact with ozone.

Ozone Hole and its Causes

The Antarctic ozone hole is an area of the Antarctic stratosphere in which the recent ozone levels have dropped to as low as 33 percent of their pre-1975 values. The ozone hole occurs during the Antarctic spring, from September to early December, as strong westerly winds start to circulate around the continent and create an atmospheric container. Within this polar vortex, over 50 percent of the lower stratospheric ozone is destroyed during the Antarctic spring.

As explained above, the primary cause of ozone depletion is the presence of chlorine-containing source gases (primarily CFCs and related halocarbons). In the presence of UV light, these gases dissociate, releasing chlorine atoms, which then go on to catalyze ozone destruction. The Cl-catalyzed ozone depletion can take place in the gas phase, but it is dramatically enhanced in the presence of polar stratospheric clouds (PSCs).

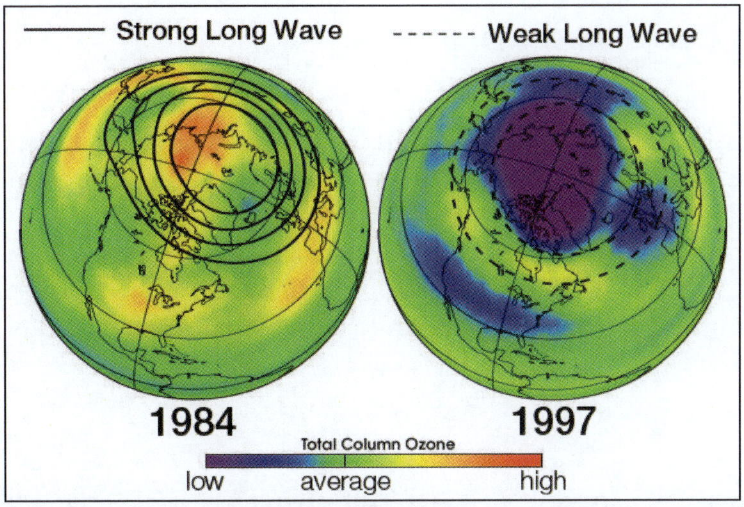

Ozone hole in North America during 1984 (abnormally warm reducing ozone depletion) and 1997 (abnormally cold resulting in increased seasonal depletion).

These polar stratospheric clouds form during winter, in the extreme cold. Polar winters are dark, consisting of three months without solar radiation (sunlight). The lack of sunlight contributes to a decrease in temperature and the polar vortex traps and chills air. Temperatures hover around or below −80 °C. These low temperatures form cloud particles. There are three types of PSC clouds—nitric acid trihydrate clouds, slowly cooling water-ice clouds, and rapid cooling water-ice (nacreous) clouds—provide surfaces for chemical reactions whose products will, in the spring lead to ozone destruction.

The photochemical processes involved are complex but well understood. The key observation is that, ordinarily, most of the chlorine in the stratosphere resides in "reservoir" compounds, primarily chlorine nitrate ($ClONO_2$) as well as stable end products such as HCl. The formation of end products essentially remove Cl from the ozone depletion process. The former sequester Cl, which can be later made available via absorption of light at shorter wavelengths than 400 nm. During the Antarctic winter and spring, however, reactions on the surface of the polar stratospheric cloud particles convert these "reservoir" compounds into reactive free radicals (Cl and ClO). The process by which the clouds remove NO_2 from the stratosphere by converting it to nitric acid in the PSC particles, which then are lost by sedimentation is called denitrification. This prevents newly formed ClO from being converted back into $ClONO_2$.

The role of sunlight in ozone depletion is the reason why the Antarctic ozone depletion is greatest during spring. During winter, even though PSCs are at their most abundant, there is no light over the pole to drive chemical reactions. During the spring, however, the sun comes out, providing energy to drive photochemical reactions and melt the polar stratospheric clouds, releasing considerable ClO, which drives the hole mechanism. Further warming temperatures near the end of spring break up the vortex around mid-December. As warm, ozone and NO_2-rich air flows in from lower latitudes, the PSCs are destroyed, the enhanced ozone depletion process shuts down, and the ozone hole closes.

Most of the ozone that is destroyed is in the lower stratosphere, in contrast to the much smaller ozone depletion through homogeneous gas phase reactions, which occurs primarily in the upper stratosphere.

Consequences of Ozone Layer Depletion

Since the ozone layer absorbs UVB ultraviolet light from the sun, ozone layer depletion increases surface UVB levels (all else equal), which could lead to damage, including increase in skin cancer. This was the reason for the Montreal Protocol. Although decreases in stratospheric ozone are well-tied to CFCs and to increases in surface UVB, there is no direct observational evidence linking ozone depletion to higher incidence of skin cancer and eye damage in human beings. This is partly because UVA, which has also been implicated in some forms of skin cancer, is not absorbed by ozone, and because it is nearly impossible to control statistics for lifestyle changes over time.

Increased UV

Ozone, while a minority constituent in Earth's atmosphere, is responsible for most of the absorption of UVB radiation. The amount of UVB radiation that penetrates through the ozone layer decreases exponentially with the slant-path thickness and density of the layer. When stratospheric ozone levels decrease, higher levels of UVB reach the Earth's surface. UV-driven phenolic formation in tree rings has dated the start of ozone depletion in northern latitudes to the late 1700s.

In October 2008, the Ecuadorian Space Agency published a report called HIPERION. The study used ground instruments in Ecuador and the last 28 years' data from 12 satellites of several countries, and found that the UV radiation reaching equatorial latitudes was far greater than expected, with the UV Index climbing as high as 24 in Quito; the WHO considers 11 as an extreme index and a great risk to health. The report concluded that depleted ozone levels around the mid-latitudes of the planet are already endangering large populations in these areas. Later, the CONIDA, the Peruvian Space Agency, published its own study, which yielded almost the same findings as the Ecuadorian study.

Biological Effects

The main public concern regarding the ozone hole has been the effects of increased surface UV radiation on human health. So far, ozone depletion in most locations has been typically a few percent and, as noted above, no direct evidence of health damage is available in most latitudes. If the high levels of depletion seen in the ozone hole were to be common across the globe, the effects could be substantially more dramatic. As the ozone hole over Antarctica has in some instances grown so large as to affect parts of Australia, New Zealand, Chile, Argentina, and South Africa, environmentalists have been concerned that the increase in surface UV could be significant.

Ozone depletion would magnify all of the effects of UV on human health, both positive (including production of vitamin D) and negative (including sunburn, skin cancer, and cataracts). In addition, increased surface UV leads to increased tropospheric ozone, which is a health risk to humans.

Basal and Squamous Cell Carcinomas

The most common forms of skin cancer in humans, basal and squamous cell carcinomas, have been strongly linked to UVB exposure. The mechanism by which UVB induces these cancers is well understood—absorption of UVB radiation causes the pyrimidine bases in the DNA molecule to form dimers, resulting in transcription errors when the DNA replicates. These cancers are relatively mild and rarely fatal, although the treatment of squamous cell carcinoma sometimes requires extensive reconstructive surgery. By combining epidemiological data with results of animal studies, scientists have estimated that every one percent decrease in long-term stratospheric ozone would increase the incidence of these cancers by two percent.

Malignant Melanoma

Another form of skin cancer, malignant melanoma, is much less common but far more dangerous, being lethal in about 15–20 percent of the cases diagnosed. The relationship between malignant melanoma and ultraviolet exposure is not yet fully understood, but it appears that both UVB and UVA are involved. Because of this uncertainty, it is difficult to estimate the effect of ozone depletion on melanoma incidence. One study showed that a 10 percent increase in UVB radiation was associated with a 19 percent increase in melanomas for men and 16 percent for women. A study of people in Punta Arenas, at the southern tip of Chile, showed a 56 percent increase in melanoma and a 46 percent increase in nonmelanoma skin cancer over a period of seven years, along with decreased ozone and increased UVB levels.

Cortical Cataracts

Epidemiological studies suggest an association between ocular cortical cataracts and UVB exposure, using crude approximations of exposure and various cataract assessment techniques. A detailed assessment of ocular exposure to UVB was carried out in a study on Chesapeake Bay Watermen, where increases in average annual ocular exposure were associated with increasing risk of cortical opacity. In this highly exposed group of predominantly white males, the evidence linking cortical opacities to sunlight exposure was the strongest to date. Based on these results, ozone depletion is predicted to cause hundreds of thousands of additional cataracts by 2050.

Increased Tropospheric Ozone

Increased surface UV leads to increased tropospheric ozone. Ground-level ozone is generally recognized to be a health risk, as ozone is toxic due to its strong oxidant properties. The risks are particularly high for young children, the elderly, and those with asthma or other respiratory difficulties. At this time, ozone at ground level is produced mainly by the action of UV radiation on combustion gases from vehicle exhausts.

Increased Production of Vitamin D

Vitamin D is produced in the skin by ultraviolet light. Thus, higher UVB exposure raises human vitamin D in those deficient in it. Recent research (primarily since the Montreal Protocol) shows that many humans have less than optimal vitamin D levels. In particular, in the U.S. population, the lowest quarter of vitamin D (<17.8 ng/ml) were found using information from the National

Health and Nutrition Examination Survey to be associated with an increase in all-cause mortality in the general population. While blood level of vitamin D in excess of 100 ng/ml appear to raise blood calcium excessively and to be associated with higher mortality, the body has mechanisms that prevent sunlight from producing vitamin D in excess of the body's requirements.

Effects on Animals

A November 2010 report by scientists at the Institute of Zoology in London found that whales off the coast of California have shown a sharp rise in sun damage, and these scientists "fear that the thinning ozone layer is to blame". The study photographed and took skin biopsies from over 150 whales in the Gulf of California and found "widespread evidence of epidermal damage commonly associated with acute and severe sunburn", having cells that form when the DNA is damaged by UV radiation. The findings suggest "rising UV levels as a result of ozone depletion are to blame for the observed skin damage, in the same way that human skin cancer rates have been on the increase in recent decades".

Effects on Crops

An increase of UV radiation would be expected to affect crops. A number of economically important species of plants, such as rice, depend on cyanobacteria residing on their roots for the retention of nitrogen. Cyanobacteria are sensitive to UV radiation and would be affected by its increase. "Despite mechanisms to reduce or repair the effects of increased ultraviolet radiation, plants have a limited ability to adapt to increased levels of UVB, therefore plant growth can be directly affected by UVB radiation".

GLOBAL DIMMING

Dozens of fires burning on the surface (red dots) and a thick pall of smoke and haze (greyish pixels) filling the skies overhead in Eastern China. Smoke, pollution and other air particles are linked to global dimming. Photo taken by MODIS aboard NASA's Aqua satellite.

Global dimming is the gradual reduction in the amount of global direct irradiance at the Earth's surface that was observed for several decades after the start of systematic measurements in the

1950s. The effect varies by location, but worldwide it has been estimated to be of the order of a 4% reduction over the three decades from 1960–1990. However, after discounting an anomaly caused by the eruption of Mount Pinatubo in 1991, a very slight reversal in the overall trend has been observed.

Global dimming is thought to have been caused by an increase in particulates such as sulfate aerosols in the atmosphere due to human action. It has interfered with the hydrological cycle by reducing evaporation and may have reduced rainfall in some areas. Global dimming also creates a cooling effect that may have partially counteracted the effect of greenhouse gases on global warming.

It is thought that global dimming is probably due to the increased presence of aerosol particles in Earth's atmosphere, caused by pollution, dust, or volcanic eruptions. Aerosols and other particulates absorb solar energy and reflect sunlight back into space. The pollutants can also become nuclei for cloud droplets. Water droplets in clouds coalesce around the particles. Increased pollution causes more particulates and thereby creates clouds consisting of a greater number of smaller droplets (that is, the same amount of water is spread over more droplets). The smaller droplets make clouds more reflective, so that more incoming sunlight is reflected back into space and less reaches the Earth's surface. This same effect also reflects radiation from below, trapping it in the lower atmosphere. In models, these smaller droplets also decrease rainfall.

Clouds intercept both heat from the sun and heat radiated from the Earth. Their effects are complex and vary in time, location, and altitude. Usually during the daytime the interception of sunlight predominates, giving a cooling effect; however, at night the re-radiation of heat to the Earth slows the Earth's heat loss.

Probable Causes

NASA photograph showing aircraft contrails and natural clouds. The temporary disappearance of contrails over North America due to plane groundings after the September 11, 2001 attacks, and the resulting increase in diurnal temperature range gave empirical evidence of the effect of thin ice clouds at the Earth's surface.

The incomplete combustion of fossil fuels (such as diesel) and wood releases black carbon into the air. Though black carbon, most of which is soot, is an extremely small component of air pollution at land surface levels, the phenomenon has a significant heating effect on the atmosphere at altitudes above two kilometers (6,562 ft). Also, it dims the surface of the ocean by absorbing solar radiation.

Experiments in the Maldives (comparing the atmosphere over the northern and southern islands) in the 1990s showed that the effect of macroscopic pollutants in the atmosphere at that time (blown south from India) caused about a 10% reduction in sunlight reaching the surface in the area under the Asian brown cloud — a much greater reduction than expected from the presence of the particles themselves. Prior to the research being undertaken, predictions were of a 0.5–1% effect from particulate matter; the variation from prediction may be explained by cloud formation with the particles acting as the focus for droplet creation.

The phenomenon underlying global dimming may also have regional effects. While most of the earth has warmed, the regions that are downwind from major sources of air pollution (specifically sulfur dioxide emissions) have generally cooled. This may explain the cooling of the eastern United States relative to the warming western part.

However some research shows that black carbon will increase global warming, being second only to CO_2. They believe that soot will absorb solar energy and transport it to other areas such as the Himalayas where glacial melting occurs. It can also darken Arctic ice reducing reflectivity and increasing absorption of solar radiation.

Airborne volcanic ash can reflect the Sun's rays back into space and thereby contribute to cooling the planet. Dips in earth temperatures have been observed after large volcano eruptions such as Mount Agung in Bali that erupted in 1963, El Chichon (Mexico) 1983, Ruiz (Colombia) 1985, and Pinatubo (Philippines) 1991. But even for major eruptions, the ash clouds remain only for relatively short periods. It has also been theorized that today's rapid climate change may increase volcanic activity.

Contrails and Clouds

A study theorized that aircraft contrails (also called vapor trails) are implicated in regional cooling, but the constant flow of air traffic previously meant that this could not be tested. The near-total shutdown of civil air traffic during the three days following the September 11, 2001 attacks afforded a unique opportunity in which to observe the climate of the United States absent from the effect of contrails. During this period, an increase in diurnal temperature variation of 1.1 °C (1.8 °F) was observed in some parts of the U.S., i.e. aircraft contrails may have been raising nighttime temperatures and/or lowering daytime temperatures by much more than previously thought. However, a follow-up study attributed cloud cover to the temperature change. The authors wrote, "The variations in high cloud cover, including contrails and contrail-induced cirrus clouds, contribute weakly to the changes in the diurnal temperature range, which is governed primarily by lower altitude clouds, winds, and humidity".

Recent Reversal of the Trend

Wild using measurements over land, report brightening since 1990, and Pinker found that slight dimming continued over land while brightening occurred over the ocean. Hence, over the land surface, Wild and Pinker disagree. A 2007 NASA sponsored satellite-based study sheds light on the puzzling observations by other scientists that the amount of sunlight reaching Earth's surface had been steadily declining in recent decades, began to reverse around 1990. This switch from a "global dimming" trend to a "brightening" trend happened just as global aerosol levels started to decline.

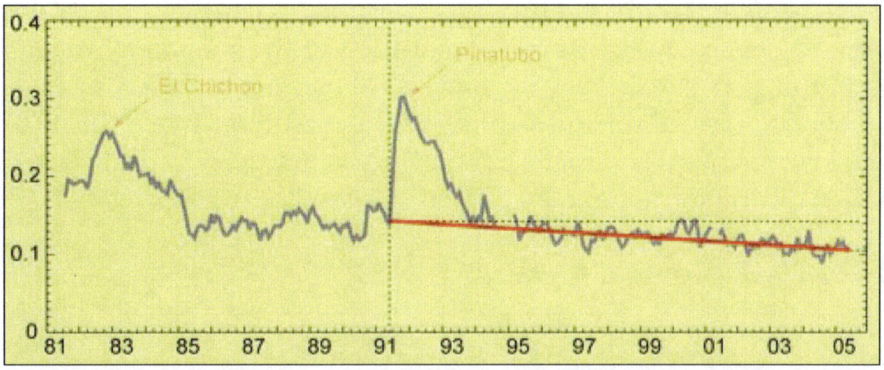

Sun-blocking aerosols around the world steadily declined (red line) since the 1991 eruption of Mount Pinatubo, according to satellite estimates. Credit: Michael Mishchenko, NASA.

It is likely that at least some of this change, particularly over Europe, is due to decreases in airborne pollution. Most governments of developed nations have taken steps to reduce aerosols released into the atmosphere, which helps reduce global dimming.

Sulfate aerosols have declined significantly since 1970 with the Clean Air Act in the United States and similar policies in Europe. The Clean Air Act was strengthened in 1977 and 1990. According to the EPA, from 1970 to 2005, total emissions of the six principal air pollutants, including PM's, dropped by 53% in the US. In 1975, the masked effects of trapped greenhouse gases finally started to emerge and have dominated ever since.

The Baseline Surface Radiation Network (BSRN) has been collecting surface measurements. BSRN was started in the early 1990s and updated the archives in this time. Analysis of recent data reveals that the surface of the planet has brightened by about 4% in the past decade. The brightening trend is corroborated by other data, including satellite analyses.

Relationship to Global Warming

Some scientists now consider that the effects of global dimming have masked the effect of global warming to some extent and that resolving global dimming may therefore lead to increases in predictions of future temperature rise. According to Beate Liepert, "We lived in a global warming plus a global dimming world and now we are taking out global dimming. So we end up with the global warming world, which will be much worse than we thought it will be, much hotter." The magnitude of this masking effect is one of the central problems in current climate change with significant implications for future climate changes and policy responses to global warming.

Interactions between the two theories for climate modification have also been studied, as global warming and global dimming are neither mutually exclusive nor contradictory. In a paper published on March 8, 2005 in the American Geophysical Union's Geophysical Research Letters, a research team led by Anastasia Romanou of Columbia University's Department of Applied Physics and Mathematics, New York, also showed that the apparently opposing forces of global warming and global dimming can occur at the same time. Global dimming interacts with global warming by blocking sunlight that would otherwise cause evaporation and the particulates bind to water droplets. Water vapor is the major greenhouse gas. On the other hand, global dimming is affected by evaporation and rain. Rain has the effect of clearing out polluted skies.

Brown clouds have been found to amplify global warming according to Veerabhadran Ramanathan, an atmospheric chemist at the Scripps Institution of Oceanography in La Jolla, California. "The conventional thinking is that brown clouds have masked as much as 50 percent of global warming by greenhouse gases through so-called global dimming. While this is true globally, this study reveals that over southern and eastern Asia, the soot particles in the brown clouds are in fact amplifying the atmospheric warming trend caused by greenhouse gases by as much as 50 percent".

Relationship to Hydrological Cycle

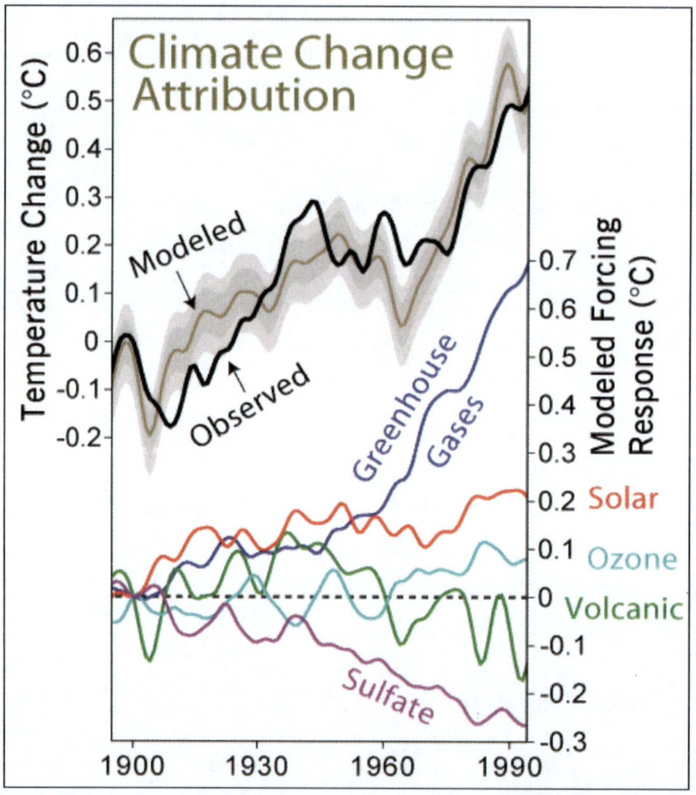

This figure shows the level of agreement between a climate model driven by five factors and the historical temperature record. The negative component identified as "sulfate" is associated with the aerosol emissions blamed for global dimming.

Pollution produced by humans may be seriously weakening the Earth's water cycle — reducing rainfall and threatening fresh water supplies. A 2001 study by researchers at the Scripps Institution of Oceanography suggests that tiny particles of soot and other pollutants have a significant effect on the hydrological cycle. According to Veerabhadran Ramanathan, "the energy for the hydrological cycle comes from sunlight. As sunlight heats the ocean, water escapes into the atmosphere and falls out as rain. So as aerosols cut down sunlight by large amounts, they may be spinning down the hydrological cycle of the planet".

Large scale changes in weather patterns may also have been caused by global dimming. Climate modelers speculatively suggest that this reduction in solar radiation at the surface may have led to the failure of the monsoon in sub-Saharan Africa during the 1970s and 1980s, together with the associated famines such as the Sahel drought, caused by Northern hemisphere pollution cooling the Atlantic. Because of this, the Tropical rain belt may not have risen to its northern latitudes, thus causing an absence of seasonal rains. This claim is not universally accepted and is very difficult to test. However a 2009 Chinese study of 50 years of continuous data found that, though most parts of eastern China saw no significant change in the amount of water held by the atmosphere, light rains had decreased. The researchers then modeled the effect of aerosols and also concluded the overall effect was that water drops in polluted cases are up to 50 percent smaller than in pristine skies. They concluded smaller size impedes the formation of rain clouds, and the falling of light rain is beneficial for agriculture. This was a different effect than reducing solar irradiance, but still a direct result from the presence of aerosols.

The 2001 study by researchers at the Scripps Institution of Oceanography concluded that the imbalance between global dimming and global warming at the surface leads to weaker turbulent heat fluxes to the atmosphere. This means globally reduced evaporation and hence precipitation occur in a dimmer and warmer world, which could ultimately lead to a more humid atmosphere in which it rains less.

A natural form of large scale environmental dimming effect on the development of tropical cyclones originates from Sahara desert dust, when the drifting sand and mineral particles laden air moves over the Atlantic Ocean. The particles reflect and absorb sunlight, less sun rays reach Earth surface layers, thus resulting in cooler water and land surface temperatures, and also less cloud formation, subsequently dampening the development of hurricanes.

Possible use to Mitigate Global Warming

Some scientists have suggested using aerosols to stave off the effects of global warming as an emergency geoengineering measure. In 1974, Mikhail Budyko suggested that if global warming became a problem, the planet could be cooled by burning sulfur in the stratosphere, which would create a haze. An increase in planetary albedo of just 0.5 percent is sufficient to halve the effect of a CO_2 doubling.

The simplest solution would be to simply emit more sulfates, which would end up in troposphere - the lowest part of the atmosphere. If this were done, Earth would still face many problems, such as:

- Using sulfates causes environmental problems such as acid rain.
- Using carbon black causes human health problems.
- Dimming causes ecological problems such as changes in evaporation and rainfall patterns.
- Droughts and increased rainfall cause problems for agriculture.
- Aerosol has a relatively short lifetime.

The solution advocated is transporting sulfates into the next higher layer of the atmosphere - stratosphere. Aerosols in the stratosphere last years instead of weeks - so only a relatively smaller (though still large) amount of sulfate emissions would be necessary, and side effects would be less. This would require developing an efficient way to transport large amounts of gases into stratosphere, many of which have been proposed though none are known to be effective or economically viable.

MILANKOVITCH CYCLES

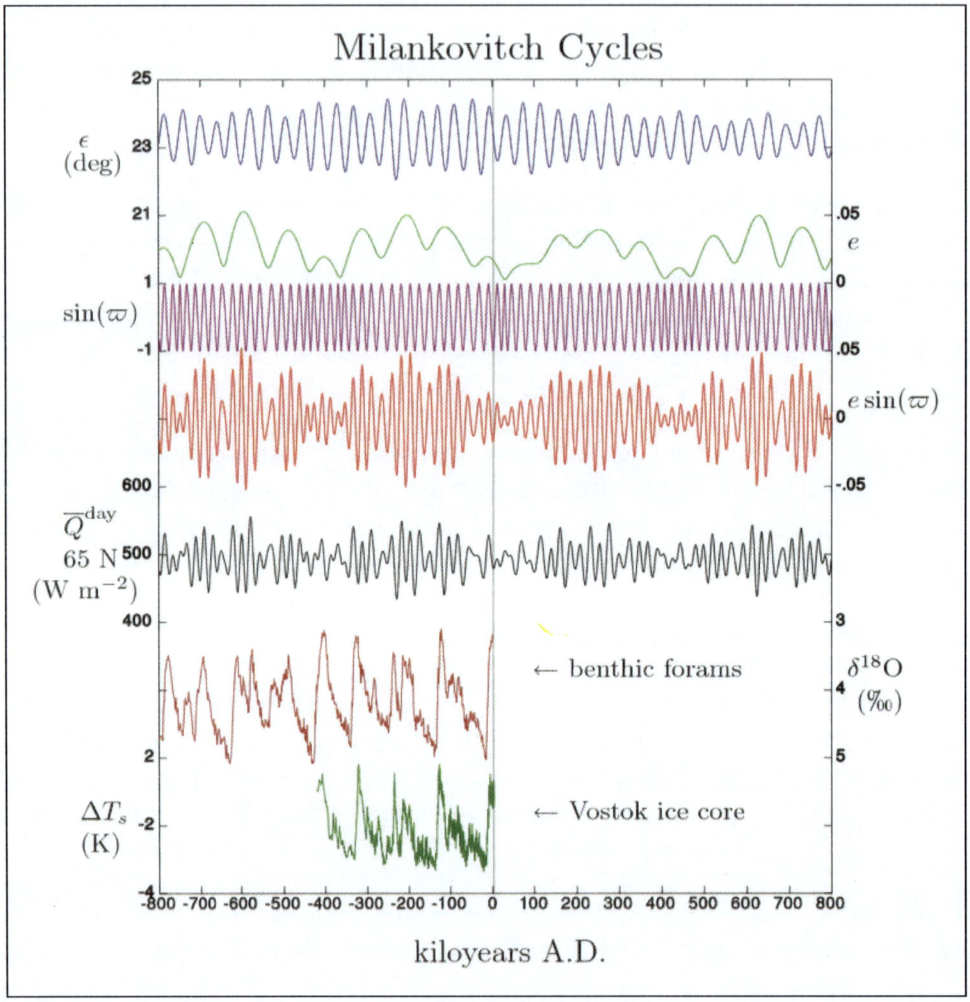

Past and future Milankovitch cycles. VSOP allows prediction of past and future orbital parameters with great accuracy.

The graphic shows variations in these five orbital elements:

- Obliquity (axial tilt) (ε).
- Eccentricity (*e*).
- Longitude of perihelion (sin(ϖ)).
- Precession index (*e* sin(ϖ)), which together with obliquity, controls the seasonal cycle of insolation.

■ Calculated daily-averaged insolation at the top of the atmosphere,

($\bar{Q}^{day}$), on the day of the summer solstice at 65° N latitude.

ƚ Data from cores of ocean sediment and Antarctic ice are two distinct proxies for global sea levels and temperatures of the past:

■ Benthic forams

■ *Vostok ice core*

The vertical gray line shows current conditions (at year 2000 A.D.)

Milankovitch cycles describe the collective effects of changes in the Earth's movements on its climate over thousands of years. The term is named for Serbian geophysicist and astronomer Milutin Milanković. In the 1920s, he hypothesized that variations in eccentricity, axial tilt, and precession of the Earth's orbit resulted in cyclical variation in the solar radiation reaching the Earth, and that this orbital forcing strongly influenced climatic patterns on Earth.

Similar astronomical hypotheses had been advanced in the 19th century by Joseph Adhemar, James Croll and others, but verification was difficult because there was no reliably dated evidence, and because it was unclear which periods were important.

Now, materials on Earth that have been unchanged for millennia (obtained via ice, rock, and deep ocean cores) are being studied to indicate the history of Earth's climate. Though they are consistent with the Milankovitch hypothesis, there are still several observations that the hypothesis does not explain.

Earth's Movements

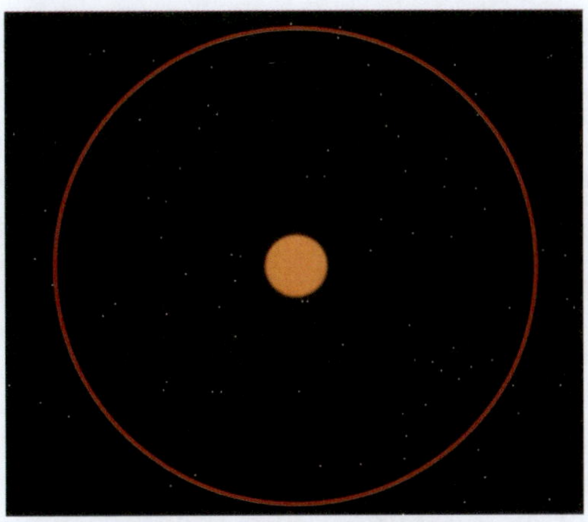

Circular orbit, no eccentricity.

The Earth's rotation around its axis, and revolution around the Sun, evolve over time due to gravitational interactions with other bodies in the solar system. The variations are complex, but a few cycles are dominant.

The Earth's orbit varies between nearly circular and mildly elliptical (its eccentricity varies). When the orbit is more elongated, there is more variation in the distance between the Earth and the Sun, and in the amount of solar radiation, at different times in the year. In addition, the rotational tilt of the Earth (its obliquity) changes slightly. A greater tilt makes the seasons more extreme. Finally, the direction in the fixed stars pointed to by the Earth's axis changes (axial precession), while the Earth's elliptical orbit around the Sun rotates (apsidal precession). The combined effect is that proximity to the Sun occurs during different astronomical seasons.

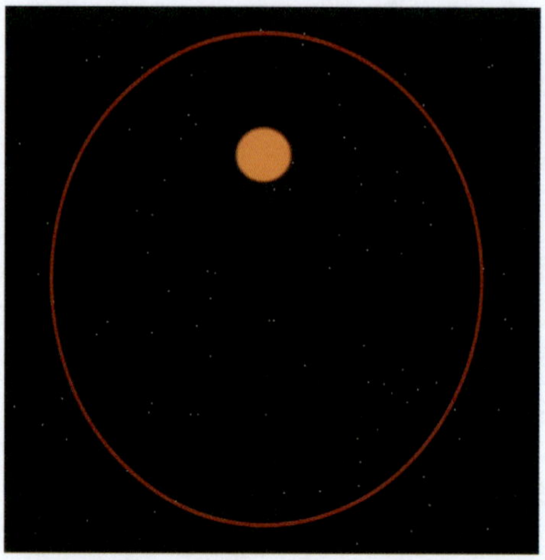

Orbit with 0.5 eccentricity, exaggerated for illustration; Earth's orbit is only slightly eccentric.

Milankovitch studied changes in these movements of the Earth, which alter the amount and location of solar radiation reaching the Earth. This is known as solar forcing (an example of radiative forcing). Milankovitch emphasized the changes experienced at 65° north due to the great amount of land at that latitude. Land masses change temperature more quickly than oceans, because of the mixing of surface and deep water and the fact that soil has a lower volumetric heat capacity than water.

Orbital Shape

The Earth's orbit approximates an ellipse. Eccentricity measures the departure of this ellipse from circularity. The shape of the Earth's orbit varies between nearly circular (with the lowest eccentricity of 0.000055) and mildly elliptical (highest eccentricity of 0.0679) Its geometric or logarithmic mean is 0.0019. The major component of these variations occurs with a period of 413,000 years (eccentricity variation of ±0.012). Other components have 95,000-year and 125,000-year cycles (with a beat period of 400,000 years). They loosely combine into a 100,000-year cycle (variation of −0.03 to +0.02). The present eccentricity is 0.017 and decreasing.

Eccentricity varies primarily due to the gravitational pull of Jupiter and Saturn. However, the semi-major axis of the orbital ellipse remains unchanged; according to perturbation theory, which computes the evolution of the orbit, the semi-major axis is invariant. The orbital period (the length of a sidereal year) is also invariant, because according to Kepler's third law, it is determined by the semi-major axis.

Effect on Temperature

The semi-major axis is a constant. Therefore, when Earth's orbit becomes more eccentric, the semi-minor axis shortens. This increases the magnitude of seasonal changes.

Season durations				
Year	Northern Hemisphere	Southern Hemisphere	Date: GMT	Season duration
2005	Winter solstice	Summer solstice	21 December 2005 18:35	88.99 days
2006	Spring equinox	Autumn equinox	20 March 2006 18:26	92.75 days
2006	Summer solstice	Winter solstice	21 June 2006 12:26	93.65 days
2006	Autumn equinox	Spring equinox	23 September 2006 4:03	89.85 days
2006	Winter solstice	Summer solstice	22 December 2006 0:22	88.99 days
2007	Spring equinox	Autumn equinox	21 March 2007 0:07	92.75 days
2007	Summer solstice	Winter solstice	21 June 2007 18:06	93.66 days
2007	Autumn equinox	Spring equinox	23 September 2007 9:51	89.85 days
2007	Winter solstice	Summer solstice	22 December 2007 06:08	

The relative increase in solar irradiation at closest approach to the Sun (perihelion) compared to the irradiation at the furthest distance (aphelion) is slightly larger than four times the eccentricity. For Earth's current orbital eccentricity, incoming solar radiation varies by about 6.8%, while the distance from the Sun currently varies by only 3.4% (5.1 million km). Perihelion presently occurs around January 3, while aphelion is around July 4. When the orbit is at its most eccentric, the amount of solar radiation at perihelion will be about 23% more than at aphelion. However, the Earth's eccentricity is always so small that the variation in solar irradiation is a minor factor in seasonal climate variation, compared to axial tilt and even compared to the relative ease of heating the larger land masses of the northern hemisphere.

Effect on Lengths of Seasons

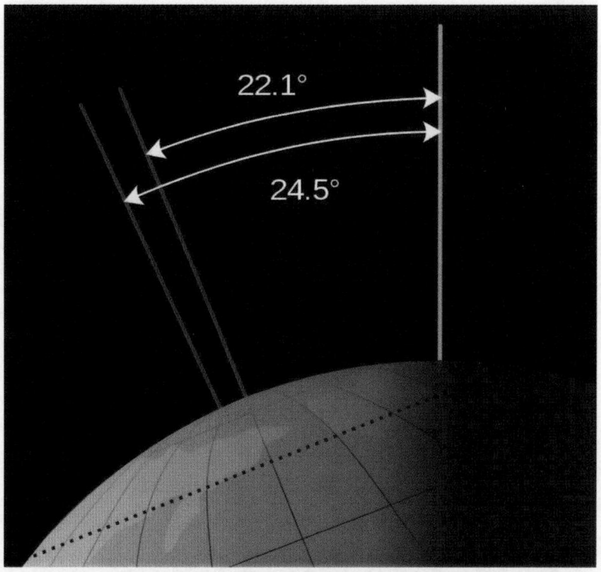

22.1–24.5° range of Earth's obliquity.

The seasons are quadrants of the Earth's orbit, marked by the two solstices and the two equinoxes. Kepler's second law states that a body in orbit traces equal areas over equal times; its orbital velocity is highest around perihelion and lowest around aphelion. The Earth spends less time near perihelion and more time near aphelion. This means that the lengths of the seasons vary.

Perihelion currently occurs around January 3, so the Earth's greater velocity shortens winter and autumn in the northern hemisphere. Summer in the northern hemisphere is 4.66 days longer than winter, and spring is 2.9 days longer than autumn.

Greater eccentricity increases the variation in the Earth's orbital velocity. However, currently, the Earth's orbit is becoming less eccentric (more nearly circular). This will make the seasons more similar in length.

Axial Tilt

The angle of the Earth's axial tilt with respect to the orbital plane (the obliquity of the ecliptic) varies between 22.1° and 24.5°, over a cycle of about 41,000 years. The current tilt is 23.44°, roughly halfway between its extreme values. The tilt last reached its maximum in 8,700 BCE. It is now in the decreasing phase of its cycle, and will reach its minimum around the year 11,800 CE.

Increased tilt increases the amplitude of the seasonal cycle in insolation, providing more solar radiation in each hemisphere's summer and less in winter. However, these effects are not uniform everywhere on the Earth's surface. Increased tilt increases the total annual solar radiation at higher latitudes, and decreases the total closer to the equator.

The current trend of decreasing tilt, by itself, will promote milder seasons (warmer winters and colder summers), as well as an overall cooling trend. Because most of the planet's snow and ice lies at high latitude, decreasing tilt may encourage the onset of an ice age for two reasons: There is less overall summer insolation, and also less insolation at higher latitudes, which melts less of the previous winter's snow and ice.

Axial Precession

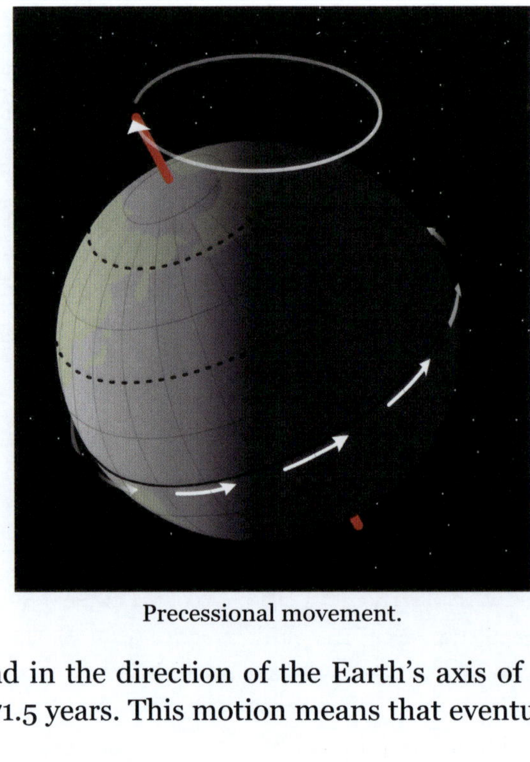

Precessional movement.

Axial precession is the trend in the direction of the Earth's axis of rotation relative to the fixed stars, with a period of 25,771.5 years. This motion means that eventually Polaris will no longer be the north pole star. It is caused by the tidal forces exerted by the Sun and the Moon on the solid Earth; both contribute roughly equally to this effect.

Planets orbiting the Sun follow elliptical (oval) orbits that rotate gradually over time (apsidal precession). The eccentricity of this ellipse, as well as the rate of precession, is exaggerated for visualization.

Currently, perihelion occurs during the southern hemisphere's summer. This means that solar radiation due to (1) axial tilt inclining the southern hemisphere toward the Sun and (2) the Earth's proximity to the Sun, both reach maximum during the summer and both reach minimum during the winter. Their effects on heating are additive, which means that seasonal variation in irradiation of the southern hemisphere is more extreme. In the northern hemisphere, these two factors reach maximum at opposite times of the year: The north is tilted toward the Sun when the Earth is furthest from the Sun. The two effects work in opposite directions, resulting in less extreme variations in insolation.

In about 13,000 years, the north pole will be tilted toward the Sun when the Earth is at perihelion. Axial tilt and orbital eccentricity will both contribute their maximum increase in solar radiation during the northern hemisphere's summer. Axial precession will promote more extreme variation in irradiation of the northern hemisphere and less extreme variation in the south.

When the Earth's axis is aligned such that aphelion and perihelion occur near the equinoxes, axial tilt will not be aligned with or against eccentricity.

Apsidal Precession

In addition, the orbital ellipse itself precesses in space, in an irregular fashion, completing a full cycle every 112,000 years relative to the fixed stars. Apsidal precession occurs in the plane of the ecliptic and alters the orientation of the Earth's orbit relative to the ecliptic. This happens primarily as a result of interactions with Jupiter and Saturn. Smaller contributions are also made by the sun's oblateness and by the effects of general relativity that are well known for Mercury.

Apsidal precession combines with the 25,771.5 year cycle of axial precession to vary the position in the year that the Earth reaches perihelion. Apsidal precession shortens this period to 23,000 years on average (varying between 20,800 and 29,000 years).

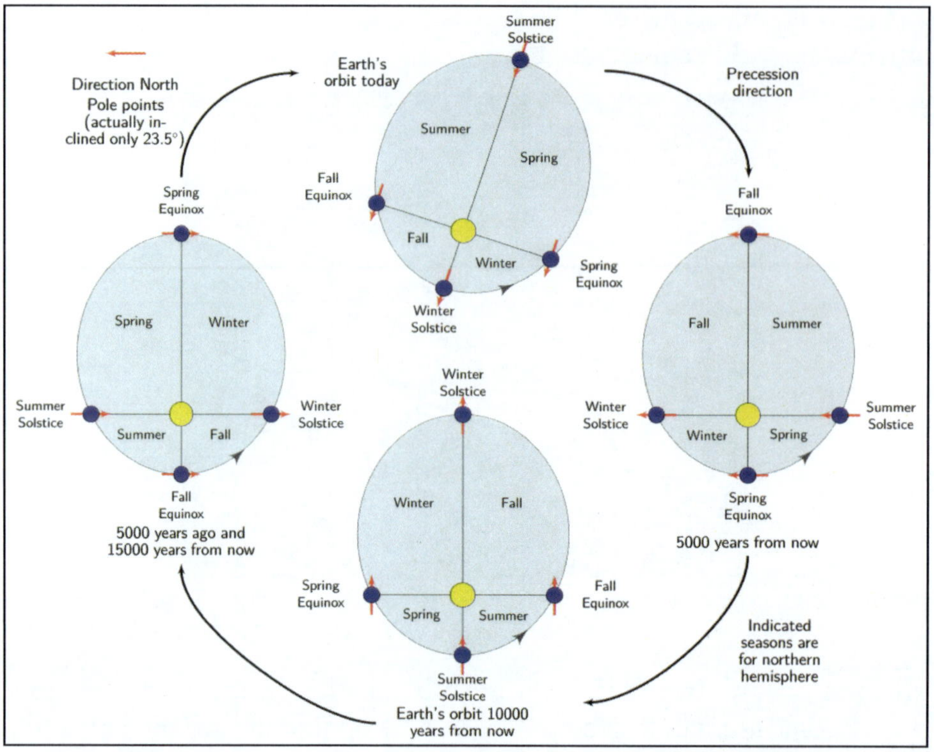

Effects of precession on the seasons (using the Northern Hemisphere terms).

As the orientation of Earth's orbit changes, each season will gradually start earlier in the year. Precession means the Earth's nonuniform motion will affect different seasons. Winter, for instance, will be in a different section of the orbit. When the Earth's apsides are aligned with the equinoxes, the length of spring and summer combined will equal that of autumn and winter. When they are aligned with the solstices, the difference in the length of these seasons will be greatest.

Orbital Inclination

The inclination of Earth's orbit drifts up and down relative to its present orbit. This three-dimensional movement is known as "precession of the ecliptic" or "planetary precession". Earth's current inclination relative to the invariable plane (the plane that represents the angular momentum of the Solar System, approximately the orbital plane of Jupiter) is 1.57°.

Milankovitch did not study apsidal precession. It was discovered more recently and measured, relative to Earth's orbit, to have a period of about 70,000 years. However, when measured independently of Earth's orbit, but relative to the invariable plane, precession has a period of about 100,000 years. This period is very similar to the 100,000-year eccentricity period. Both periods closely match the 100,000-year pattern of glacial events.

Theory Constraints

Artifacts taken from the Earth have been studied to infer the cycles of past climate. A study of the chronology of Antarctic ice cores using oxygen-nitrogen ratios in air bubbles trapped in the ice, which appear to respond directly to the local insolation, concluded that the climatic response documented in the ice cores was driven by northern hemisphere insolation as proposed by the

Milankovitch hypothesis. Analysis of deep-ocean cores, analysis of lake depths, and a seminal paper by Hays, Imbrie, and Shackleton provide additional validation through physical artifacts. Climate records contained in a 1,700 ft (520 m) core of rock drilled in Arizona show a pattern synchronized with Earth's eccentricity, and cores drilled in New England match it, going back 215 million years.

The nature of sediments can vary in a cyclic fashion, and these cycles can be displayed in the sedimentary record. Here, cycles can be observed in the colouration and resistance of different strata.

100,000-Year Issue

Of all the orbital cycles, Milankovitch believed that obliquity had the greatest effect on climate, and that it did so by varying the summer insolation in northern high latitudes. Therefore, he deduced a 41,000-year period for ice ages. However, subsequent research has shown that ice age cycles of the Quaternary glaciation over the last million years have been at a period of 100,000 years, which matches the eccentricity cycle.

Various explanations for this discrepancy have been proposed, including frequency modulation or various feedbacks (from carbon dioxide, cosmic rays, or from ice sheet dynamics). Some models can reproduce the 100,000-year cycles as a result of non-linear interactions between small changes in the Earth's orbit and internal oscillations of the climate system.

Jung-Eun Lee of Brown University proposes that precession changes the amount of energy that Earth absorbs, because the southern hemisphere's greater ability to grow sea ice reflects more energy away from Earth. Moreover, Lee says, "Precession only matters when eccentricity is large. That's why we see a stronger 100,000-year pace than a 21,000-year pace".

Some have argued that the length of the climate record is insufficient to establish a statistically significant relationship between climate and eccentricity variations.

Transition Changes

In fact, from 1–3 million years ago, climate cycles did match the 41,000-year cycle in obliquity. After 1 million years ago, the Mid-Pleistocene Transition (MPT) occurred with switch to the

100,000-year cycle matching eccentricity. The *transition problem* refers to the need to explain what changed 1 million years ago. The MPT can now be reproduced in numerical simulations that include a decreasing trend in carbon dioxide and glacially induced removal of regolith.

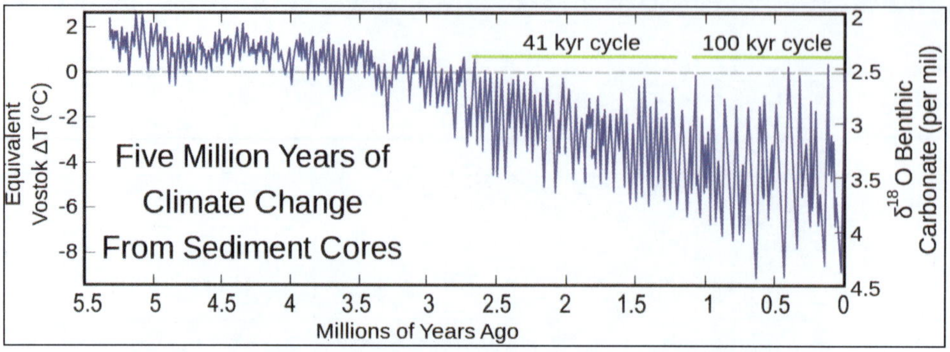

Variations of cycle times, curves determined from ocean sediments.

Interpretation of Unsplit Peak Variances

Even the well-dated climate records of the last million years do not exactly match the shape of the eccentricity curve. Eccentricity has component cycles of 95,000 and 125,000 years. However, some researchers say the records do not show these peaks, but only show a single cycle of 100,000 years.

Unsynched Stage 5 Observation

Deep-sea core samples show that the interglacial interval known as marine isotope stage 5 began 130,000 years ago. This is 10,000 years before the solar forcing that the Milankovitch hypothesis predicts.

Predicted Effects Mystery

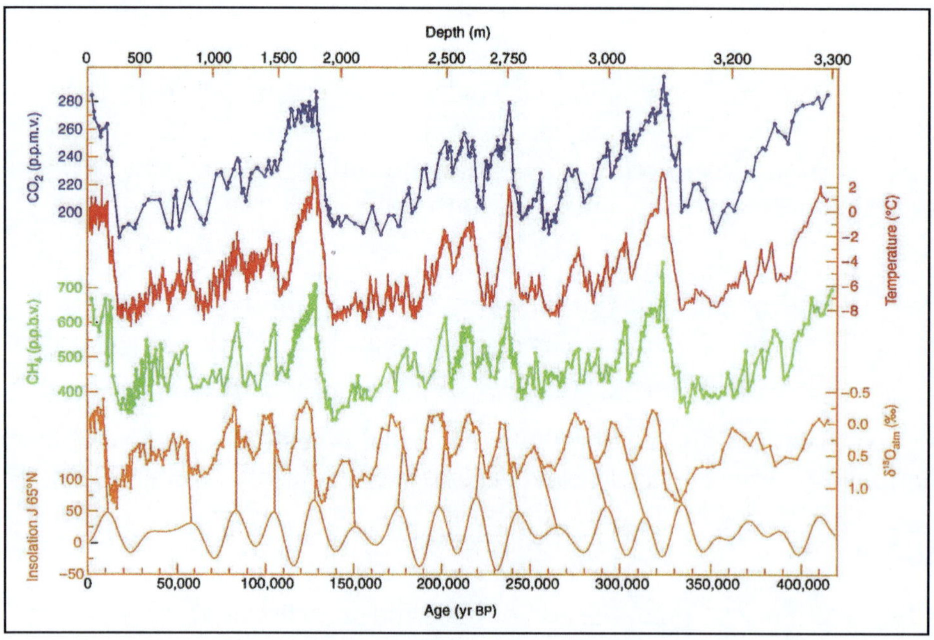

420,000 years of ice core data from Vostok, Antarctica research station, with more recent times on the left.

Artifacts show that the variation in Earth's climate is much more extreme than the variation in the intensity of solar radiation calculated as the Earth's orbit evolves. If orbital forcing causes climate change, science needs to explain why the observed effect is amplified compared to the theoretical effect.

Some climate systems exhibit amplification (positive feedback) and damping responses (negative feedback). An example of amplification would be if, with the land masses around 65° north covered in year-round ice, solar energy were reflected away. Amplification would mean that an ice age induces changes that impede orbital forcing from ending the ice age.

The Earth's current orbital inclination is 1.57°. Earth presently moves through the invariable plane around January 9 and July 9. At these times, there is an increase in meteors and noctilucent clouds. If this is because there is a disk of dust and debris in the invariable plane, then when the Earth's orbital inclination is near 0° and it is orbiting through this dust, materials could be accreted into the atmosphere. This process could explain the narrowness of the 100,000-year climate cycle.

Present and Future Conditions

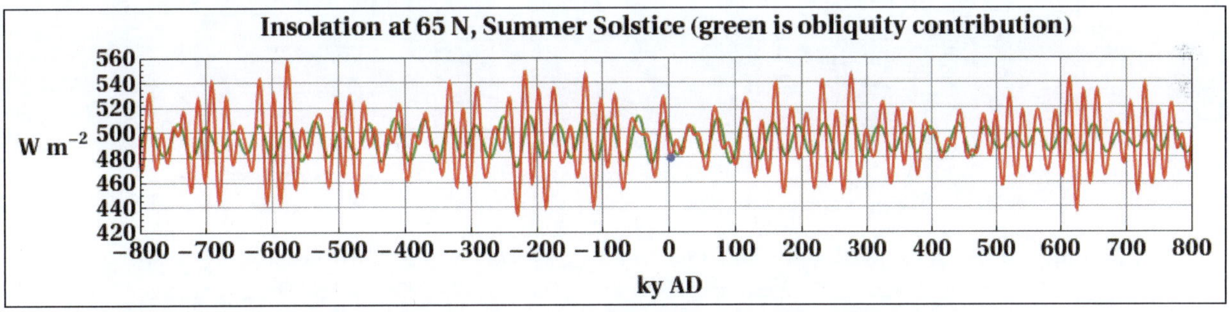

Past and future of daily average insolation at top of the atmosphere on the day of the summer solstice, at 65 N latitude. The green curve is with eccentricity *e* hypothetically set to 0. The red curve uses the actual (predicted) value of *e*. Blue dot is current conditions, at 2 ky A.D.

Since orbital variations are predictable, any model that relates orbital variations to climate can be run forward to predict future climate, with two caveats: the mechanism by which orbital forcing influences climate is not definitive; and non-orbital effects can be important (for example, Human impact on the environment principally increases in greenhouse gases result in a warmer climate).

An often-cited 1980 orbital model by Imbrie predicted "the long-term cooling trend that began some 6,000 years ago will continue for the next 23,000 years." More recent work suggests that orbital variations should gradually increase 65° N summer insolation over the next 25,000 years. Earth's orbit will become less eccentric for about the next 100,000 years, so changes in this insolation will be dominated by changes in obliquity, and should not decline enough to permit a new glacial period in the next 50,000 years.

GLACIAL PERIOD

A glacial period (alternatively glacial or glaciation) is an interval of time (thousands of years) within an ice age that is marked by colder temperatures and glacier advances. Interglacials, on the other hand, are periods of warmer climate between glacial periods. The last glacial period ended about 15,000 years ago. The Holocene epoch is the current interglacial. A time with no glaciers on Earth is considered a greenhouse climate state.

Quaternary Ice Age

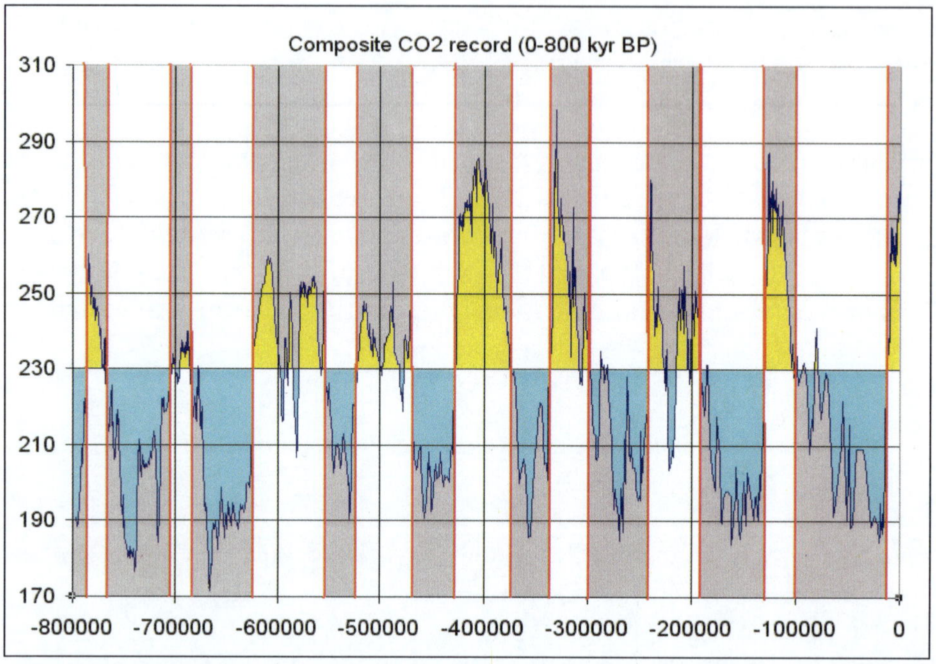

Glacial and interglacial cycles as represented by atmospheric CO_2, measured from ice core samples going back 800,000 years. The stage names are part of the North American and the European Alpine subdivisions. The correlation between both subdivisions is tentative.

Within the Quaternary (about 2.6 Ma to present), there have been a number of glacials and interglacials.

Penultimate Glacial Period

The Penultimate Glacial Period (PGP) is the glacial period that occurred before the Last Glacial Period. It began ~194,000 years ago, and ended 135,000 years ago with the beginning of the Eemian interglacial.

Last Glacial Period

The last glacial period was the most recent glacial period within the Quaternary Ice Age, occurring in the Pleistocene epoch, which began about 110,000 years ago and ended about 15,000 years ago. The glaciations that occurred during this glacial period covered many areas of the Northern Hemisphere and have different names, depending on their geographic distributions: *Wisconsin* (in North America), *Devensian* (in Great Britain), *Midlandian* (in Ireland), *Würm* (in the Alps), *Weichsel* (in northern central Europe), *Dali* (in East China), *Beiye* (in North China), *Taibai* (in Shaanxi) *Luojishan* (in Southwest Sichuan), *Zagunao* (in Northwest Sichuan), *Tianchi* (in Tianshan Mountains) *Qomolangma* (in Himalaya), and *Llanquihue* (in Chile). The glacial advance reached its maximum extent about 26,500 BP. In Europe, the ice sheet reached Northern Germany. In the last 650,000 years, there were, on average, seven cycles of glacial advance and retreat.

Next Glacial Period

Since orbital variations are predictable, computer models that relate orbital variations to climate can predict future climate possibilities. Work by Berger and Loutre suggests that the current warm climate may last another 50,000 years. The amount of heat trapping (greenhouse) gases being emitted into Earth's Oceans and atmosphere may delay the next glacial period by an additional 50,000 years.

ORBITAL FORCING

Orbital forcing involves the redistribution of incoming solar energy, both latitudinally and seasonally. Thus, there are differential effects on the climate system that can lead to circulation changes, and there may be different responses to the forcing in the Northern and Southern Hemispheres. Changes in solar irradiance (energy emitted by the Sun) might be expected to affect all parts of the Earth equally. However, this is not so, because the response to solar irradiance forcing is amplified regionally, as a result of feedbacks and interactions within the atmosphere.

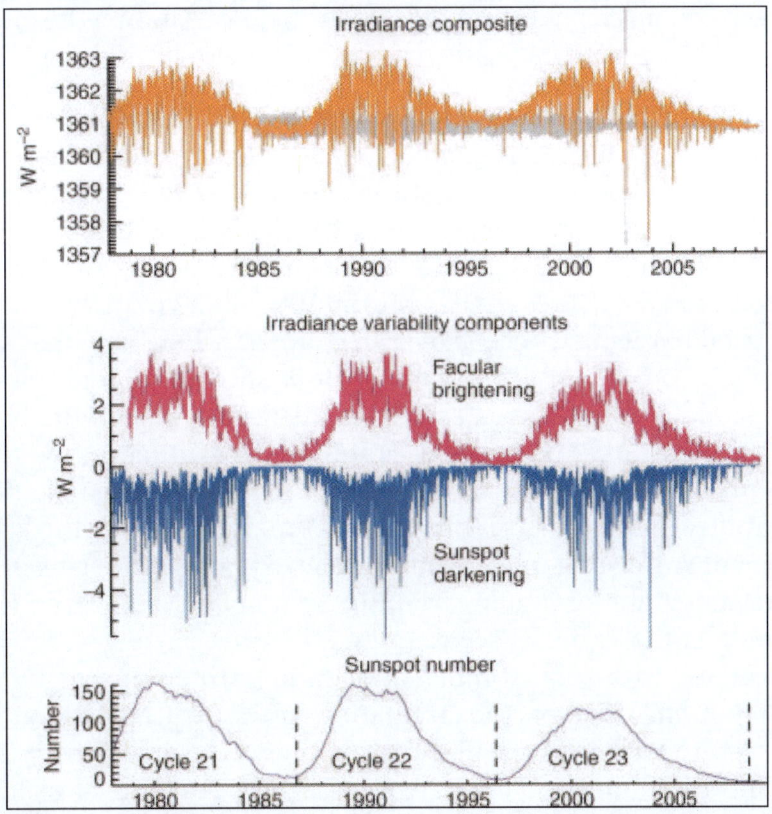

Satellite measurements of solar faculae and sunspots for the last three solar activity cycles. The faculae increase energy emissions from the Sun, while sunspots reduce them. The net result is a 0.1% increase in total solar irradiance between the minimum and maximum of the solar cycles (top). Unfortunately, longer measurements of these phenomena do not exist, and so, there is considerable uncertainty about how irradiance changed in the past.

For many years, it was assumed (based on measurements in dry, high-altitude locations) that total irradiance did not vary, at least not on interannual to decadal scales—hence the term "solar constant" was coined to describe the energy that is intercepted by the atmosphere when the Sun is overhead. Satellite measurements over the last ~ 25 years tell a different story—total solar irradiance, TSI (incoming radiation integrated over all wavelengths) varies by ~ 0.08% over a Schwabe solar cycle (average length of ~ 11 years), with maximum values at times of maximum solar activity (when there are many sunspots and bright solar faculae) (Lean, 2010). Sunspots reduce solar irradiance and faculae increase it, so the overall changes during a solar cycle reflect the net effect of these two phenomena, with faculae outweighing sunspots during solar maxima. Furthermore, irradiance changes at very short (ultraviolet) wavelengths and vary even more over a solar cycle. Such changes have significance because an increase in UV radiation causes more ozone (O_3) to be produced in the upper stratosphere. Ozone absorbs radiation (at UV wavelengths of 200-340 nm) so heating rates in the upper atmosphere are increased during times of enhanced solar activity. This then affects stratospheric winds (strengthening stratospheric easterlies), which can in turn influence surface climate via dynamical linkages between the stratosphere and the troposphere. Model simulations of these effects indicate that there is a poleward shift of the tropospheric westerly jet and a poleward extension of the Hadley circulation, by ~ 70 km from solar minimum to solar maximum, in the summer hemisphere. Although such changes are small, if irradiance changes in the past were larger and more persistent than solar cycle variability, the effects may have been quite significant.

How much has irradiance changed over longer timescales? Satellite measurements are too short to shed light on longer-term irradiance changes, so these must be inferred from other lines of evidence. Sunspot observations began in the early seventeenth century, but observations of solar faculae are much shorter, so the difficulty is how to estimate changes in total irradiance when the only long-term data are from sunspot numbers . Of particular interest is the period from ~ 1645 to 1715 (the Maunder Minimum) when virtually no sunspots were seen. Many studies have tried to reconstruct the overall change in TSI since the Maunder Minimum, with estimates varying from < 0.1% to 0.24%, with presumably larger changes in UV radiation. Numerous GCM simulations have used these estimates to examine how solar irradiance may have affected temperatures over recent centuries. Notwithstanding the uncertainties in estimating the magnitude of solar forcing, these studies generally conclude that much of the low-frequency variability in Northern Hemisphere temperatures over the last millennium (prior to the onset of global anthropogenic effects) can be explained in terms of solar and volcanic forcing, though internal climate system variability may mask the overall (large-scale) warming and cooling effects at the regional scale. In fact, distinct patterns of regional temperature change may be associated with solar forcing, as seen in both empirical and modeling studies, due to complex interactions between the circulation in the stratosphere and the troposphere, and feedbacks between the atmosphere and the ocean circulation. By varying solar irradiance forcing in a GCM, Shindell simulated cooler winter temperatures over eastern North America and western Europe (by 1-2 °C), compared to a century later when solar irradiance was higher, changes that are consistent with the paleoclimatic evidence for the late Maunder Minimum. Such a pattern is characteristic of a shift in the North Atlantic Oscillation (NAO) toward lower index conditions, in which the pressure gradient between Iceland and the Azores is reduced, leading to less advection of warm, moist air from the Atlantic into western Europe and cooler temperatures over Eurasia.

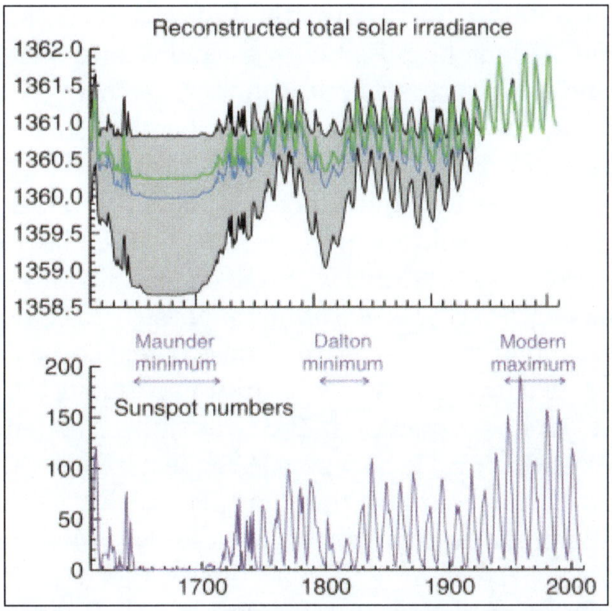

Sunspot numbers recorded since the early 1600s, showing the strong 11-year Schwabe cycle and longer-term variations, from near zero during the Maunder Minimum to over 150 in recent decades. In the upper diagram, estimates of the changes in total solar irradiance (TSI) are given, based on different models of solar activity. The latest estimate for solar minimum during the last solar cycle is 1360 W m^{-2}.

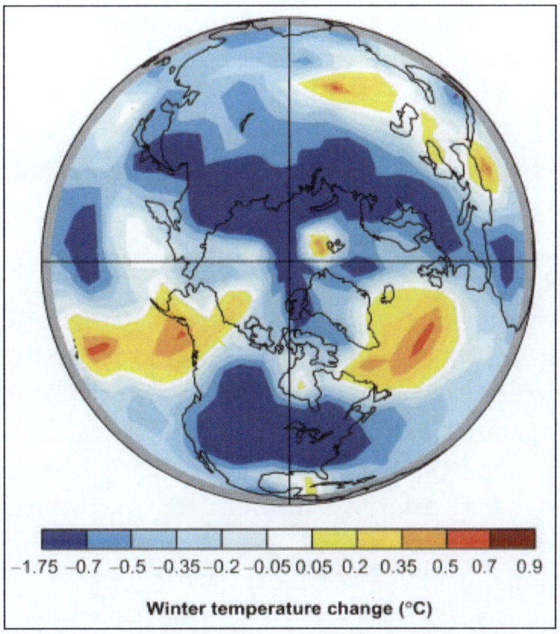

The change in winter surface air temperature between 1680 and 1780, based on a GCM simulation in which solar irradiance was increased by 0.32 W m^{-2} (~ 0.1%) between those times. The GCM had an interactive stratospheric chemistry component so that UV effects on ozone could be included. The pattern of change is similar to that associated with a negative phase of the North Atlantic Oscillation.

If solar irradiance changed by ~ 0.1-0.24% over the last 350 years, how much change occurred over longer periods of time? It may be possible to estimate long-term changes in solar activity from changes in cosmogenic isotopes preserved in natural archives. Cosmic rays intercepted by the upper atmosphere produce cosmogenic isotopes—such as ^{10}Be and ^{14}C—which eventually enter the terrestrial environment at the Earth's surface. During times of high solar activity, the flux of cosmic rays to the atmosphere is reduced, leading to a reduction in the production rate of these isotopes. Thus, variations in cosmogenic isotopes are inversely related to solar activity. If we make the assumption that solar activity is correlated with TSI changes (a fairly big assumption, though it is supported by observations in the recent instrumental period), long-term changes in ^{14}C (seen as departures from expected age, in tree rings) or ^{10}Be (in ice cores) could be used as an index of solar irradiance changes over time. However, there are questions about the role of atmospheric processes in transferring ^{10}Be from its production in the stratosphere to deposition sites at the Earth's surface and about the rate at which radiocarbon was sequestered in the deep ocean (due, for example, to thermohaline circulation changes) that may also have affected atmospheric concentrations in radiocarbon over time. Furthermore, other factors have also affected the production rate of cosmogenic isotopes over the Holocene, and these must be accounted for in order to isolate the effects of solar variability. In particular, geomagnetic field variations have had a large impact on production rates (a weaker field being associated with higher production levels), and variations of the Earth's magnetic field are not well understood. To examine some of these questions, Bard used a box model of ocean carbon variations driven by ^{10}Be variations measured in an ice core to assess whether ^{14}C variations over the last 1000 years had been affected by ocean circulation changes.

On this time scale, such effects appear to have been minimal, suggesting that ^{14}C can be used to assess solar variability over the millennium and perhaps longer. A similar result was found by Beer for the last 4000 years. Thus, many studies have compared paleoclimatic reconstructions with ^{14}C anomalies, as an implicit proxy for solar variability. In this way, precipitation variability across the tropics, in northern South America and Yucatan in east Africa and the Arabian Peninsula, and in China, has been correlated with solar activity variations, represented by ^{14}C anomalies (though the total amount of variance accounted for is quite small). Solar variability has also been linked to midcontinental drought frequency on both short and long timescales, and other studies have identified potential links between solar activity variations and climate changes in the Holocene. Bond argued that the abundance of ice-rafted debris in North Atlantic sediments varied with the same frequency as ^{14}C anomalies, and Stuiver also noted the similarity between a ~ 1470-year periodicity in ^{14}C data and a similar periodicity in oxygen isotopic data from GISP2. It remains to be seen how robust these relationships are and what plausible mechanisms might link very small solar activity/irradiance variations with climate in such diverse parts of the globe. One possibility (seen in some model simulations) involves solar variations influencing the tropical Hadley circulation, which leads to changes in the distribution of clouds and precipitation anomalies, with further teleconnections to extratropical regions.

A more physically based reconstruction of total solar irradiance changes over the Holocene has been produced by Steinhilber, based on a calibration of ^{10}Be from ice cores and observed variations in the open magnetic field of the Sun. After considering estimates of how the Earth's magnetic field has changed over the Holocene, their reconstructed irradiance reveals a large number of anomalies that were comparable to the Maunder Minimum (as well as episodes of enhanced solar activity). Note, however, that a different estimate of geomagnetic field effects could shift those values. Nevertheless, their estimates indicate that TSI generally varied by less than ± 0.1% from modern levels during the Holocene, which means that solar effects on surface climate are probably brought about by strong feedbacks within the climate system. Longer-term variations have not yet been estimated.

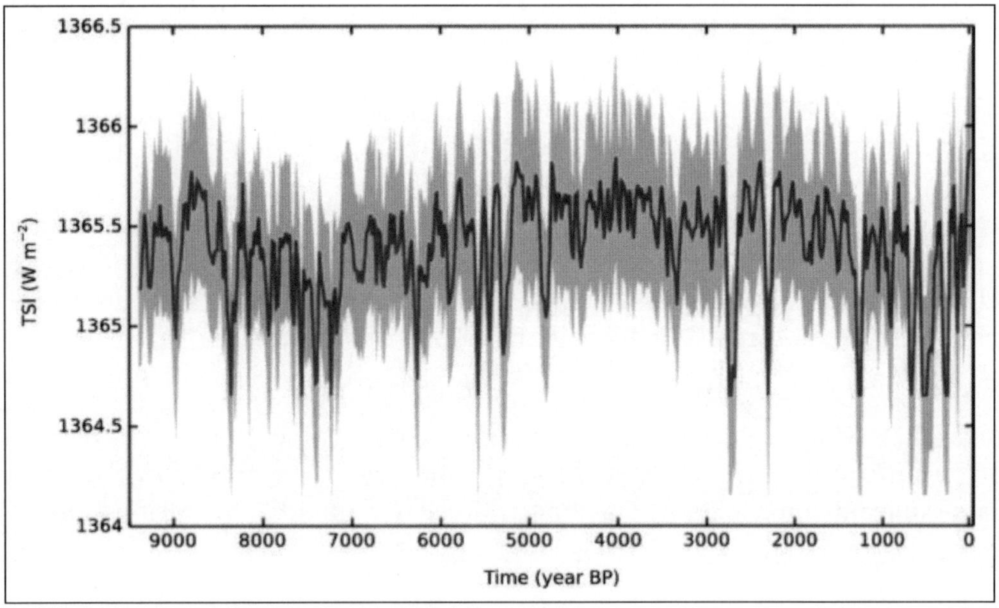

A reconstruction of total solar irradiance for the last 9300 years based on the relationship between the Sun's open magnetic field (which modulates galactic cosmic rays reaching the Earth) and ^{10}Be measured in ice cores, after taking into account changes in the Earth's magnetic field. The shaded band is the 1σ uncertainty considering the uncertainties of the calibration and of the reconstruction of the Sun's open magnetic field record.

References

- Rebecca, Lindsey (14 January 2009). "Climate and Earth's Energy Budget : Feature Articles". Earthobservatory.nasa.gov. Retrieved 3 April 2018
- Causes, climate, explore: dugreen.teri.res.in, Retrieved 1 March, 2019
- Soden, Brian J.; Held, Isaac M.; Colman, Robert; Shell, Karen M.; Kiehl, Jeffrey T.; Shields, Christine A. (2008-07-01). "Quantifying Climate Feedbacks Using Radiative Kernels". Journal of Climate. 21 (14): 3504–3520
- Orbital-forcing, earth-and-planetary-sciences, topics: sciencedirect.com, Retrieved 2 April, 2019
- Malhi, Y., et al. "Climate Change, Deforestation, and the Fate of the Amazon." Science, vol. 319, no. 5860, 11 Jan. 2008, pp. 169–172., doi:10.1126/science.1146961
- Grundmann, Reiner (2001). Transnational Environmental Policy: the ozone layer. New York: Routledge. ISBN 978-0-415-22423-9
- "Arctic ozone loss at record level". BBC News Online. October 2, 2011. Archived from the original on October 2, 2011. Retrieved October 3, 2011

3

Effects of Climate Change

Changes in the climate pose numerous adverse impacts on the environment and human life. Some of these impacts include forest fires, tropical cyclones, droughts, global warming, etc. which severely affect agriculture, ecosystem and fisheries. This chapter closely examines these key effects of climate change to provide an extensive understanding of the subject.

Climate change destabilises the Earth's temperature equilibrium and has far-reaching effects on human beings and the environment. During the course of global warming, the energy balance and thus the temperature of the earth change, due to the increased concentration of greenhouse gases, which has a significant impact on humans and the environment. The various effects of climate change are:

Temperature Will Rise: Climate models predict that Earth's global average temperate will rise in the future. For the next two decades warming of about 0.2° Celsius is projected. If we continue to emit as many, or more, greenhouse gases, this will cause more warming during the 21st Century than we saw in the 20th Century. During the 21st Century, various computer models predict that Earth's average temperature will rise between 1.8° and 4.0° Celsius (3.2° and 7.2° F). The amount of predicted warming differs depending on the model emissions scenario (how much greenhouse gas emissions it assumes for the future). The amount of predicted warming also differs between different climate models. Climate change is predicted to impact regions differently. For example, temperature increases are expected to be greater on land than over oceans and greater at high latitudes than in the tropics and mid-latitudes.

Changing Precipitation: Warmer average global temperature will cause a higher rate of evaporation, causing the water cycle to "speed up". More water vapor in the atmosphere will lead to more precipitation. According to models, global average precipitation will most likely increase by about 3-5% with a minimum increase of at least 1% and a maximum increase of about 8%. Yet, changes in precipitation will not be evenly distributed. Some locations will get more snow, others will see less rain. Some places will have wetter winters and drier summers.

Melting Snow and Ice: As the climate warms, snow and ice melt. The amount of summer melting of glaciers, ice sheets, and other snow and ice on land are predicted to be greater than the amount of winter precipitation. The amount of sea ice (frozen seawater) floating in the ocean in the Arctic and Antarctic is expected to decrease over the 21st Century too, although there is some uncertainty as to the amount of melt.

Rising Sea Level: A warmer climate causes sea level to rise via two mechanisms: (1) melting glaciers and ice sheets (ice on land) add water to the oceans, raising sea level, and (2) ocean water expands as it warms, increasing its volume and thus also raising sea level. During the 20th Century, sea level rose about 10 to 20 cm (4 to 8 inches). Thermal expansion and melting ice each contributed about half of the rise, though there is some uncertainty in the exact magnitude of the contribution from each source. By the year 2100, models predict sea level will rise between about 20 and 50 cm (8 to 20 inches) above late 20th Century levels. Thermal expansion of sea water is predicted to account for about 75% of future sea level rise according to most models.

Acidic Ocean Water: Earth's oceans are predicted to act as a buffer against climate change by taking up some of the excess heat and carbon dioxide from the atmosphere. This is good news in the short run, but more problematic in the long run. Carbon dioxide combined with seawater forms weak carbonic acid. Scientists believe this process has reduced the pH of the oceans by about 0.1 pH since pre-industrial times. Further acidification of 0.14 to 0.35 pH is expected by the year 2100. More acidic ocean water may cause problems for marine organisms.

Impacts on Ocean Currents: Large-scale ocean currents called thermohaline circulation, driven by differences in salinity and temperature, may also be disrupted as climate warms. Changes in precipitation patterns and the influx of fresh water into the oceans from melting ice can alter salinity. Changing salinity, along with rising water temperature, may disrupt the currents. In an extreme case, thermohaline circulation could be disrupted or even shut down in some parts of the ocean, which could have large effects on climate.

Changing Severe Weather: Some climate scientists believe that hurricanes, typhoons, and other tropical cyclones will (and may have begun to already) change as a result of global warming. Warm ocean surface waters provide the energy that drives these immense storms. Warmer oceans in the future are expected to cause intensification of such storms. Although there may not be more tropical cyclones worldwide in the future, some scientists believe there will be a higher proportion of the most powerful and destructive storms. Some scientists believe we are already seeing evidence for an upswing in the numbers of the most powerful storms; others are less convinced.

More Clouds: Clouds are a bit of a wild-card in global climate models. Warmer global temperatures produce faster overall evaporation rates, resulting in more water vapor in the atmosphere... and hence more clouds. Different types of clouds at different locations have different effects on climate. Some shade the Earth, cooling climate. Others enhance the greenhouse effect with their heat-trapping water vapor and droplets. Scientists expect a warmer world to be a cloudier one, but are not yet certain how the increased cloudiness will feed back into the climate system. Modeling the influence of clouds in the climate system is an area of active scientific research.

Changes to Life and the Carbon Cycle: Climate change will alter many aspects of biological systems and the global carbon cycle. Temperature changes will alter the natural ranges of many types of plants and animals, both wild and domesticated. There will also be changes to the lengths of growing seasons, geographical ranges of plants, and frost dates.

Models of the global carbon cycle suggest that the Earth system will be able to absorb less CO_2 out of the atmosphere as the climate warms, worsening the warming problem.

ABRUPT CLIMATE CHANGE

Clathrate hydrates have been identified as a possible agent for abrupt changes.

An abrupt climate change occurs when the climate system is forced to transition to a new climate state at a rate that is determined by the climate system energy-balance, and which is more rapid than the rate of change of the external forcing. Past events include the end of the Carboniferous Rainforest Collapse, Younger Dryas, Dansgaard-Oeschger events, Heinrich events and possibly also the Paleocene–Eocene Thermal Maximum. The term is also used within the context of global warming to describe sudden climate change that is detectable over the time-scale of a human lifetime, possibly as the result of feedback loops within the climate system.

Timescales of events described as 'abrupt' may vary dramatically. Changes recorded in the climate of Greenland at the end of the Younger Dryas, as measured by ice-cores, imply a sudden warming of +10 °C (+18 °F) within a timescale of a few years. Other abrupt changes are the +4 °C (+7.2 °F) on Greenland 11,270 years ago or the abrupt +6 °C (11 °F) warming 22,000 years ago on Antarctica. By contrast, the Paleocene-Eocene thermal maximum may have initiated anywhere between a few decades and several thousand years. Finally, Earth Systems models project that under ongoing greenhouse gas emissions as early as 2047, the Earth's near surface temperature could depart from the range of variability in the last 150 years, affecting over 3 billion people and most places of great species diversity on Earth.

Possible tipping elements in the climate system include regional effects of global warming, some of which had abrupt onset and may therefore be regarded as abrupt climate change. Scientists have stated, "Our synthesis of present knowledge suggests that a variety of tipping elements could reach their critical point within this century under anthropogenic climate change".

It has been postulated that teleconnections, oceanic and atmospheric processes, on different timescales, connect both hemispheres during abrupt climate change.

The IPCC states that global warming "could lead to some effects that are abrupt or irreversible".

A 2013 report from the U.S. National Research Council called for attention to the abrupt impacts of climate change, stating that even steady, gradual change in the physical climate system can have abrupt impacts elsewhere, such as in human infrastructure and ecosystems, if critical thresholds are crossed. The report emphasizes the need for an early warning system that could help society better anticipate sudden changes and emerging impacts.

Scientific understanding of abrupt climate change is generally poor. The probability of abrupt change for some climate related feedbacks may be low. Factors that may increase the probability of abrupt climate change include higher magnitudes of global warming, warming that occurs more rapidly and warming that is sustained over longer time periods.

Climate Models

Climate models are unable yet to predict abrupt climate change events, or most of the past abrupt climate shifts. A potential abrupt feedback due to thermokarst lake formations in the Arctic, in response to thawing permafrost soils, releasing additional greenhouse gas methane, is currently not accounted for in climate models.

Possible Precursor

Most abrupt climate shifts are likely due to sudden circulation shifts, analogous to a flood cutting a new river channel. The best-known examples are the several dozen shutdowns of the North Atlantic Ocean's Meridional Overturning Circulation during the last ice age, affecting climate worldwide.

- The current warming of the Arctic, the duration of the summer season, is considered abrupt and massive.
- Antarctic ozone depletion caused significant atmospheric circulation changes.
- There have also been two occasions when the Atlantic's Meridional Overturning Circulation lost a crucial safety factor. The Greenland Sea flushing at 75 °N shut down in 1978, recovering over the next decade. Then the second-largest flushing site, the Labrador Sea, shut down in 1997 for ten years. While shutdowns overlapping in time have not been seen during the 50 years of observation, previous total shutdowns had severe worldwide climate consequences.

Effects

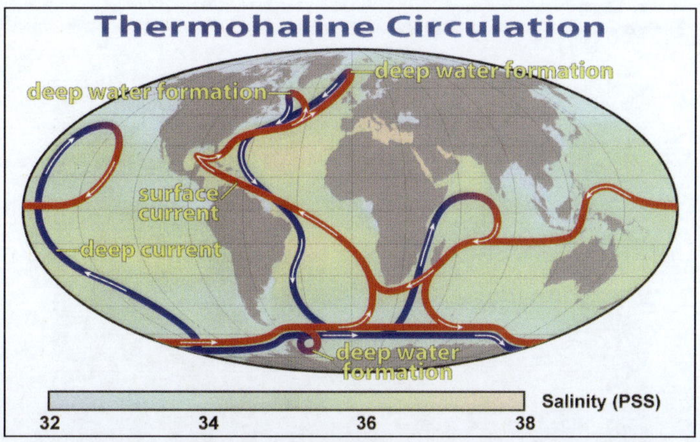

A summary of the path of the thermohaline circulation. Blue paths represent deep-water currents, and red paths represent surface currents.

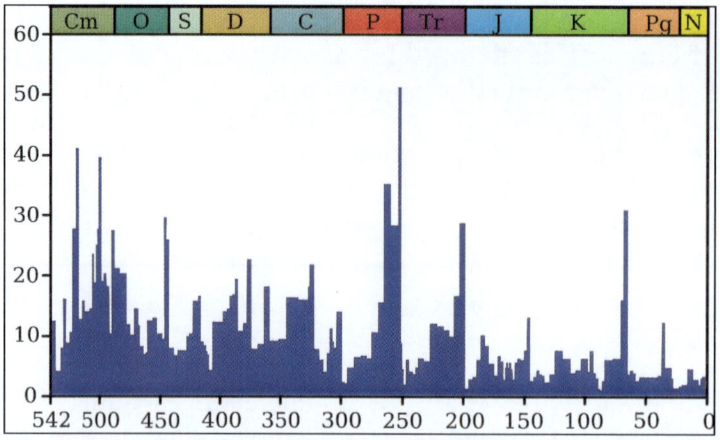

The Permian–Triassic extinction event, labelled "P-Tr" here, is the most significant extinction event in this plot for marine genera.

Abrupt climate change has likely been the cause of wide ranging and severe effects:

- Mass extinctions in the past, most notably the Permian-Triassic Extinction event (often referred to as the great dying) and the Carboniferous Rainforest Collapse, have been suggested as a consequence of abrupt climate change.
- Loss of biodiversity: without interference from abrupt climate change and other extinction events the biodiversity of this planet would continue to grow.
- Changes in ocean circulation such as:
    - Increasing frequency of El Niño events.
    - Potential disruption to the thermohaline circulation, such as that which may have occurred during the Younger Dryas event.
    - Changes to the North Atlantic oscillation.
    - Changes in Atlantic Meridional Overturning Circulation (AMOC) which could contribute to more severe weather events.

Climate Feedback Effects

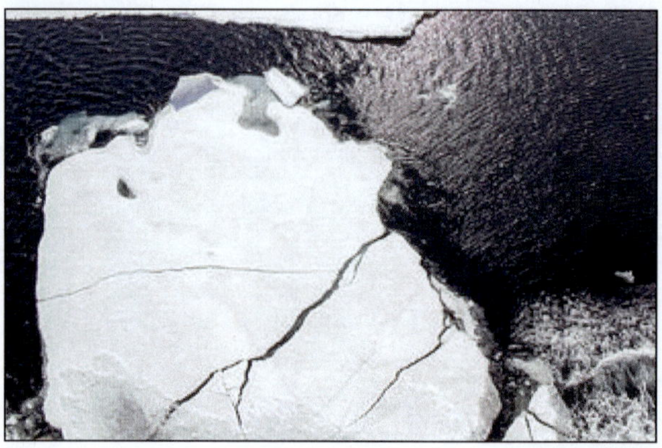
The dark ocean surface reflects only 6 percent of incoming solar radiation, instead sea ice reflects 50 to 70 percent.

One source of abrupt climate change effects is a feedback process, in which a warming event causes a change that ads to further warming. The same can apply to cooling. Example of such feedback processes are:

- Ice-albedo feedback in which the advance or retreat of ice cover alters the albedo ("whiteness") of the earth and its ability to absorb the sun's energy.
- Soil carbon feedback is the release of carbon from soils in response to global warming.
- The dying and the burning of forests by global warming.

Volcanism

Isostatic rebound in response to glacier retreat (unloading) and increased local salinity have been attributed to increased volcanic activity at the onset of the abrupt Bølling-Allerød warming. They are associated with the interval of intense volcanic activity, hinting at a interaction between climate and volcanism: enhanced short-term melting of glaciers, possibly via albedo changes from particle fallout on glacier surfaces.

PHYSICAL IMPACTS OF CLIMATE CHANGE

The change referred to may be due to natural causes but in the political context is more likely to be that due to the result of human activity, global warming.

"Climate change" refers to a change in the state of the climate that can be identified by changes in the mean and variability of its properties, and that persists for extended periods, typically decades or longer.

Global Warming

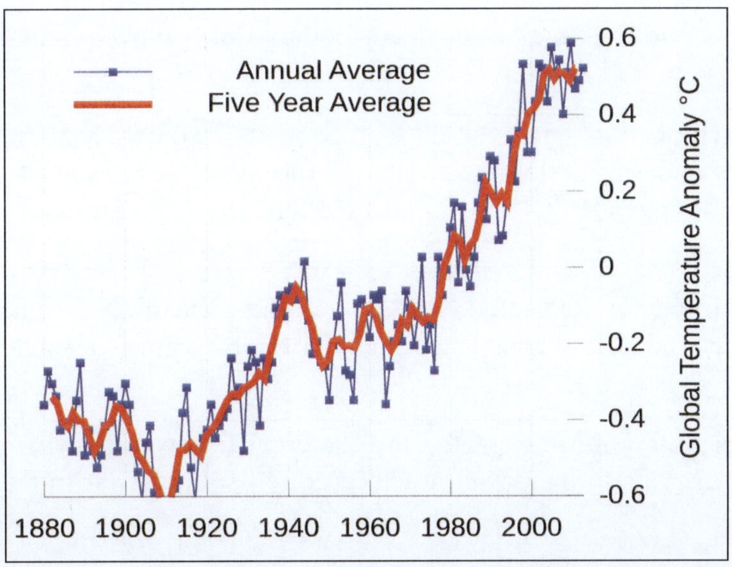

Global mean surface temperature difference from the average for 1880–2009.

Global surface temperatures have increased about 0.74 °C (plus or minus 0.18 °C) since the late-19th century, and the linear trend for the past 50 years of 0.13 °C (plus or minus 0.03 °C) per decade is nearly twice that for the past 100 years. The warming has not been globally uniform. Some areas have, in fact, cooled slightly over the last century. The recent warmth has been greatest over North America and Eurasia between 40 and 70°N. Lastly, seven of the eight warmest years on record have occurred since 2001 and the 10 warmest years have all occurred since 1995.

Effects on Weather

Increasing temperature is likely to lead to increasing precipitation but the effects on storms are less clear. Extratropical storms partly depend on the temperature gradient, which is predicted to weaken in the northern hemisphere as the polar region warms more than the rest of the hemisphere. It is possible that the Polar and Ferrel cells in one or both hemispheres will weaken and eventually disappear, which would cause the Hadley cell to cover the whole planet. This would greatly decrease the temperature gradient between the arctic and the tropics, and cause the earth to flip to a hothouse state.

Precipitation

Projected change in annual average precipitation by the end of the 21st century, based on a medium emissions scenario.

Historically, subtropical land regions have been mostly semi-arid, while most subpolar regions have had an excess of precipitation over evaporation. Future global warming is expected to be accompanied by a reduction in rainfall in the subtropics and an increase in precipitation in subpolar latitudes and some equatorial regions. In other words, regions which are dry at present will generally become even drier, while regions that are currently wet will generally become even wetter. This projection does not apply to every locale, and in some cases can be modified by local conditions. Drying is projected to be strongest near the poleward margins of the subtropics (for example, South Africa, southern Australia, the Mediterranean, and the south-western U.S.), a pattern that can be described as a poleward expansion of these semi-arid zones.

This large-scale pattern of change is a robust feature present in nearly all of the simulations conducted by the world's climate modeling groups for the 4th Assessment of the Intergovernmental Panel on Climate Change (IPCC), and is also evident in observed 20th century precipitation trends.

Changes in regional climate are expected to include greater warming over land, with most warming at high northern latitudes, and least warming over the Southern Ocean and parts of the North Atlantic Ocean.

Future changes in precipitation are expected to follow existing trends, with reduced precipitation over subtropical land areas, and increased precipitation at subpolar latitudes and some equatorial regions.

Extreme Events

Fire

Fire is a major agent for conversion of biomass and soil organic matter to CO_2. There is a large potential for future alteration in the terrestrial carbon balance through altered fire regimes. With high confidence, Schneider projected that:

- An increase in global mean temperature of about 0 to 2 °C by 2100 relative to the 1990–2000 period would result in increased fire frequency and intensity in many areas.
- An increase in the region of 2 °C or above would lead to increased frequency and intensity of fires.

Sensitivity to fires in areas that were already vulnerable has been steadily increasing. In high altitude temperate areas, increased temperature is causing snow pack to melt sooner and in greater quantities. The number of days that of higher stream flow caused by snowmelt in the Mississippi, Missouri, and Ohio rivers has been increasing in recent years. The substantial amount of snow that remains atop mountains year around is also disappearing. This leads to the surrounding densely forested areas becoming more dry and staying dry for longer periods of time. In the 1970s, the length of a fire season, which is the period of the year fires are most likely to occur, was about five months. Today, the period is usually seven months, extending into the springtime mud season. In addition, many areas are experiencing higher than normal droughts. Between 2011 and 2014, California experienced the driest period in its recorded history and more than 100 million trees died in the drought, creating areas of dead, dry wood. The decrease in rainfall is also going to increase the risk of wildfire by allowing the fire access to drier fuels. Dry foliage is more susceptible to a wildfire trigger. Wildfire specialists use foliar moisture content to determine how susceptible an area is to a wildfire. In the United States, 2015 was the most destructive year on record for wildfires, with a total of 10,125,149 total acres destroyed by fires. 2017 was the second worst year on record with 10,026,086 acres destroyed. The Thomas Fire occurred in 2017 and was the largest fire in California's history.

The increasing frequency of wildfires as a result of climate change will also lead to an increase in the amount of CO_2 in the atmosphere. This will, in turn, increase the temperature and the frequency of hot days, which will further increase fire danger. It was forecasted that double levels of CO_2, would bring a greater risk of wildfires to Australia, especially the Australian outback. All of the eight sites tested projected an increase in fire danger as a result of CO_2 level increase and all but one projected a longer fire season. The largest population center said to be affected is Alice Springs, a city deep in the Outback.

Extreme Weather

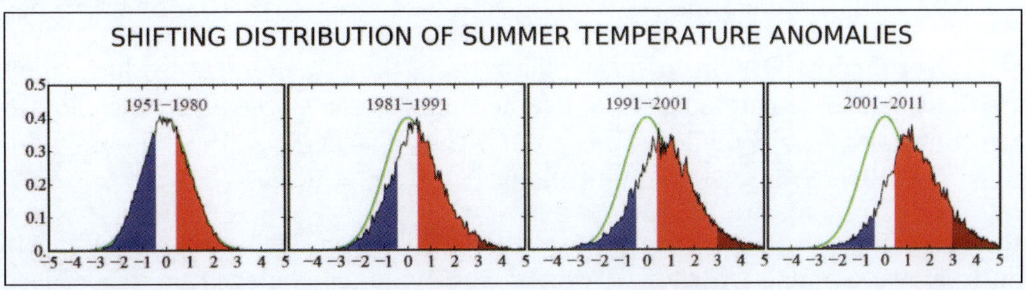

Frequency of occurrence (vertical axis) of local June–July–August temperature anomalies for Northern Hemisphere land in units of local standard deviation (horizontal axis). According to Hansen, the distribution of anomalies has shifted to the right as a consequence of global warming, meaning that unusually hot summers have become more common. This is analogous to the rolling of a die: cool summers now cover only half of one side of a six-sided die, white covers one side, red covers four sides, and an extremely hot (red-brown) anomaly covers half of one side.

IPCC projected that in the future, over most land areas, the frequency of warm spells or heat waves would very likely increase. Other likely changes are listed below:

- Increased areas will be affected by drought.
- There will be increased intense tropical cyclone activity.
- There will be increased incidences of extreme high sea level (excluding tsunamis).

Storm strength leading to extreme weather is increasing, such as the power dissipation index of hurricane intensity. Kerry Emanuel writes that hurricane power dissipation is highly correlated with temperature, reflecting global warming. However, a further study by Emanuel using current model output concluded that the increase in power dissipation in recent decades cannot be completely attributed to global warming. Hurricane modeling has produced similar results, finding that hurricanes, simulated under warmer, high-CO_2 conditions, are more intense, however, hurricane frequency will be reduced. Worldwide, the proportion of hurricanes reaching categories 4 or 5 – with wind speeds above 56 metres per second – has risen from 20% in the 1970s to 35% in the 1990s. Precipitation hitting the US from hurricanes has increased by 7% over the 20th century. The extent to which this is due to global warming as opposed to the Atlantic Multidecadal Oscillation is unclear. Some studies have found that the increase in sea surface temperature may be offset by an increase in wind shear, leading to little or no change in hurricane activity. Hoyos have linked the increasing trend in number of category 4 and 5 hurricanes for the period 1970–2004 directly to the trend in sea surface temperatures.

Thomas Knutson and Robert E. Tuleya of NOAA stated in 2004 that warming induced by greenhouse gas may lead to increasing occurrence of highly destructive category-5 storms. In 2008, Knutson found that Atlantic hurricane and tropical storm frequencies could reduce under future greenhouse-gas-induced warming. Vecchi and Soden find that wind shear, the increase of which acts to inhibit tropical cyclones, also changes in model-projections of global warming. There are projected increases of wind shear in the tropical Atlantic and East Pacific associated with the deceleration of the Walker circulation, as well as decreases of wind shear in the western and central Pacific. The study does not make claims about the net effect on Atlantic and East Pacific hurricanes of the warming and moistening atmospheres, and the model-projected increases in Atlantic wind shear.

The World Meteorological Organization explains that "though there is evidence both for and against the existence of a detectable anthropogenic signal in the tropical cyclone climate record to date, no firm conclusion can be made on this point". They also clarified that "no individual tropical cyclone can be directly attributed to climate change".

A substantially higher risk of extreme weather does not necessarily mean a noticeably greater risk of slightly-above-average weather. However, the evidence is clear that severe weather and moderate rainfall are also increasing. Increases in temperature are expected to produce more intense convection over land and a higher frequency of the most severe storms.

Using the Palmer Drought Severity Index, a 2010 study by the National Center for Atmospheric Research projects increasingly dry conditions across much of the globe in the next 30 years, possibly reaching a scale in some regions by the end of the century that has rarely, if ever, been observed in modern times.

Coumou estimated that global warming had increased the probability of local record-breaking monthly temperatures worldwide by a factor of 5. This was compared to a baseline climate in which no global warming had occurred. Using a medium global warming scenario, they project that by 2040, the number of monthly heat records globally could be more than 12 times greater than that of a scenario with no long-term warming.

Increased Evaporation

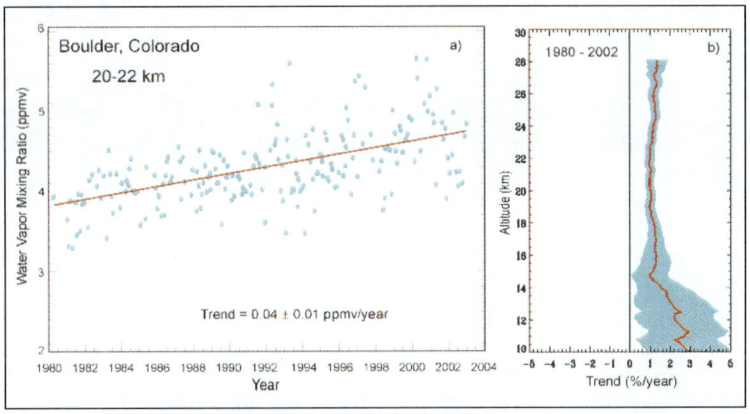

Increasing water vapor at Boulder, Colorado.

Over the course of the 20th century, evaporation rates have reduced worldwide; this is thought by many to be explained by global dimming. As the climate grows warmer and the causes of global dimming are reduced, evaporation will increase due to warmer oceans. Because the world is a closed system this will cause heavier rainfall, with more erosion. This erosion, in turn, can in vulnerable tropical areas (especially in Africa) lead to desertification. On the other hand, in other areas, increased rainfall lead to growth of forests in dry desert areas.

Scientists have found evidence that increased evaporation could result in more extreme weather as global warming progresses. The IPCC Third Annual Report says: "global average water vapor concentration and precipitation are projected to increase during the 21st century. By the second half of the 21st century, it is likely that precipitation will have increased over northern mid- to high latitudes and Antarctica in winter. At low latitudes there are both regional increases and decreases over land areas. Larger year-to-year variations in precipitation are very likely over most areas where an increase in mean precipitation is projected".

Increased Freshwater Flow

Research based on satellite observations, published in October 2010, shows an increase in the flow of freshwater into the world's oceans, partly from melting ice and partly from increased precipitation driven by an increase in global ocean evaporation. The increase in global freshwater flow, based on data from 1994 to 2006, was about 18%. Much of the increase is in areas which already experience high rainfall. One effect, as perhaps experienced in the 2010 Pakistan floods, is to overwhelm flood control infrastructure.

Regional Climate Change

General Effects

In a literature assessment, Hegerl assessed evidence for attributing observed climate change. They concluded that since the middle of the 20th century, it was likely that human influences had significantly contributed to surface temperature increases in every continent except Antarctica. A study reported on December 23, 2008, that the 10 places most affected by climate change were Darfur, the Gulf Coast, Italy, northern Europe, the Great Barrier Reef, island nations, Washington, D.C., the Northwest Passage, the Alps, and Uganda.

Northern Hemisphere

In the northern hemisphere, the southern part of the Arctic region (home to 4,000,000 people) has experienced a temperature rise of 1 °C to 3 °C (1.8 °F to 5.4 °F) over the last 50 years. Canada, Alaska and Russia are experiencing initial melting of permafrost. This may disrupt ecosystems and by increasing bacterial activity in the soil lead to these areas becoming carbon sources instead of carbon sinks. A study of changes to eastern Siberia's permafrost suggests that it is gradually disappearing in the southern regions, leading to the loss of nearly 11% of Siberia's nearly 11,000 lakes since 1971. At the same time, western Siberia is at the initial stage where melting permafrost is creating new lakes, which will eventually start disappearing as in the east. Furthermore, permafrost melting will eventually cause methane release from melting permafrost peat bogs.

Polar Regions

Anisimov assessed the literature on impacts of climate change in Polar regions. Model projections showed that Arctic terrestrial ecosystems and the active layer (the top layer of soil or rock in permafrost that is subjected to seasonal freezing and thawing) would be a small sink for carbon (i.e., net uptake of carbon) over this century. These projections were viewed as being uncertain. It was judged that increased emissions of carbon from thawing of permafrost could occur. This would lead to an amplification of warming.

Atmosphere

Temperature trends in the lower stratosphere, mid to upper troposphere, lower troposphere, and surface.

An enhanced greenhouse effect is expected to cause cooling in higher parts of the atmosphere. Cooling of the lower stratosphere (about 49,000-79,500 ft.) since 1979 is shown by both satellite Microwave sounding unit and radiosonde data, but is larger in the radiosonde data likely due to uncorrected errors in the radiosonde data.

A contraction of the thermosphere has been observed as a possible result in part due to increased carbon dioxide concentrations, the strongest cooling and contraction occurring in that layer during solar minimum. The most recent contraction in 2008–2009 was the largest such since at least 1967.

Recent evidence suggests that warming of the tropical oceans since a "tipping point" in 2000 may have acted as a negative feedback, reducing the observed warming during the 2000s (decade). As warming and evaporation above the Pacific Ocean, temperatures in the lower stratosphere near the tropopause declined due to both greenhouse gases and ozone-depleting substances, reducing water vapor levels and removing its warming effect, with vapor concentrations below 2.2 ppmv as measured by the HALOE instrument on the Upper Atmosphere Research Satellite, in the lower stratosphere of the tropics between 5°N - 5°S first being observed since 2001, although a reversal in this pattern is also likely. The water vapor in the stratosphere arrives through tall thunderstorms, while 15% of this vapor is delivered by tropical cyclones, and through chemical breakdown of methane into water vapor and carbon dioxide, both of which are greenhouse gases. The vapor is frozen out of the stratosphere as more of the water is subjected to temperatures that freeze it out of the stratosphere. Water vapor concentrations in the lower stratosphere have declined by 10% (0.4 ppmv) since 2000, reducing warming during the decade by 25%. A rapid cooling of 4 °C to 6 °C also occurred in the lower stratosphere in the mid-1990s, while the rate of ocean warming increased. During the 1990s, increased stratospheric water vapor led to a 30% increase in warming. After 2000, the sea surface temperatures of the tropical Western Pacific, where a warm pool of water exists and where temperatures are heavily influenced by ENSO, between 10°N - 10°S and 139° - 171° longitude became anti-correlated with temperatures at the tropopause in the same latitudes between 171° - 200° longitude, both measured since the early 1980s; although the correlation had been previously positive, since 2000 the SST anomalies increased while tropopause temperatures decreased. A sharp increase in average SSTs within the Western Pacific warm pool by more than 0.25 °C in 2000, which has since stabilized, occurred as the "cold point" temperature of the study area at the tropopause experienced a significant reduction. This resulted in less water vapor from tropical thunderstorms entering the stratosphere. However, prior to 2000, increases in average Western Pacific SSTs had resulted in increases in tropopause cold point temperatures.

Geophysical Systems

Biogeochemical Cycles

Climate change can have an effect on the carbon cycle in an interactive "feedback" process . A feedback exists where an initial process triggers changes in a second process that in turn influences the initial process. A positive feedback intensifies the original process, and a negative feedback reduces it. Models suggest that the interaction of the climate system and the carbon cycle is one where the feedback effect is positive.

Using the A2 SRES emissions scenario, Schneider found that this effect led to additional warming by 2100, relative to the 1990–2000 period, of 0.1 to 1.5 °C. This estimate was made with high confidence. The climate projections made in the IPCC Fourth Assessment Report of 1.1 to 6.4 °C account for this feedback effect. On the other hand, with medium confidence, Schneider commented that additional releases of GHGs were possible from permafrost, peat lands, wetlands, and large stores of marine hydrates at high latitudes.

Gas Hydrates

Gas hydrates are ice-like deposits containing a mixture of water and gas, the most common gas of which is methane. Gas hydrates are stable under high pressures and at relatively low temperatures and are found underneath the oceans and permafrost regions. Future warming at intermediate depths in the world's oceans, as predicted by climate models, will tend to destabilize gas hydrates resulting in the release of large quantities of methane. On the other hand, projected rapid sea level rise in the coming centuries associated with global warming will tend to stabilize marine gas hydrate deposits.

Carbon Cycle

Models have been used to assess the effect that climate change will have on the carbon cycle. In the Coupled Climate-Carbon Cycle Model Intercomparison Project, eleven climate models were used. Observed emissions were used in the models and future emission projections were based on the IPCC SRES A2 emissions scenario.

Unanimous agreement was found among the models that future climate change will reduce the efficiency of the land and ocean carbon cycle to absorb human-induced CO_2. As a result, a larger fraction of human-induced CO_2 will stay airborne if climate change controls the carbon cycle. By the end of the 21st century, this additional CO_2 in the atmosphere varied between 20 and 220 ppm for the two extreme models, with most models lying between 50 and 100 ppm. This additional CO_2 led to a projected increase in warming of between 0.1 and 1.5 °C.

Cryosphere

Northern Hemisphere average annual snow cover has declined in recent decades. This pattern is consistent with warmer global temperatures. Some of the largest declines have been observed in the spring and summer months.

Sea Ice

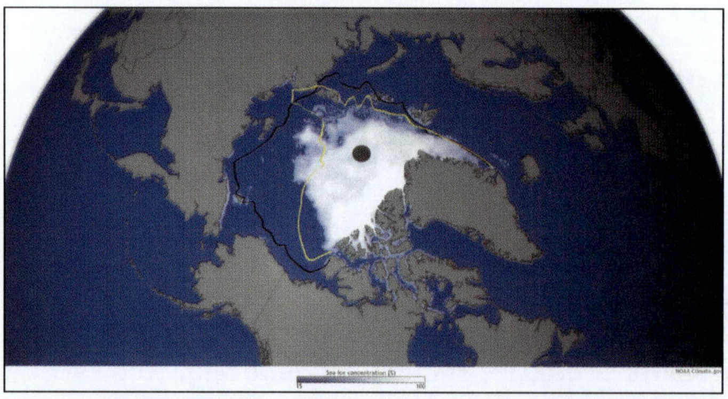

Record low Arctic sea ice extent for September 2012.

As the climate warms, snow cover and sea ice extent decrease. Large-scale measurements of sea-ice have only been possible since the satellite era, but through looking at a number of different satellite estimates, it has been determined that September Arctic sea ice has decreased between 1973 and 2007 at a rate of about -10% +/- 0.3% per decade. Sea ice extent for September for 2012 was by far the lowest on record at 3.29 million square kilometers, eclipsing the previous record low sea ice extent of 2007 by 18%. The age of the sea ice is also an important feature of the state of the sea ice cover, and for the month of March 2012, older ice has decreased from 26% of the ice cover in 1988 to 7% in 2012. Sea ice in the Antarctic has shown very little trend over the same period, or even a slight increase since 1979. Though extending the Antarctic sea-ice record back in time is more difficult due to the lack of direct observations in this part of the world.

In a literature assessment, Meehl found that model projections for the 21st century showed a reduction of sea ice in both the Arctic and Antarctic. The range of model responses was large. Projected reductions were accelerated in the Arctic. Using the high-emission A2 SRES scenario, some models projected that summer sea ice cover in the Arctic would disappear entirely by the latter part of the 21st century.

Glacier Retreat and Disappearance

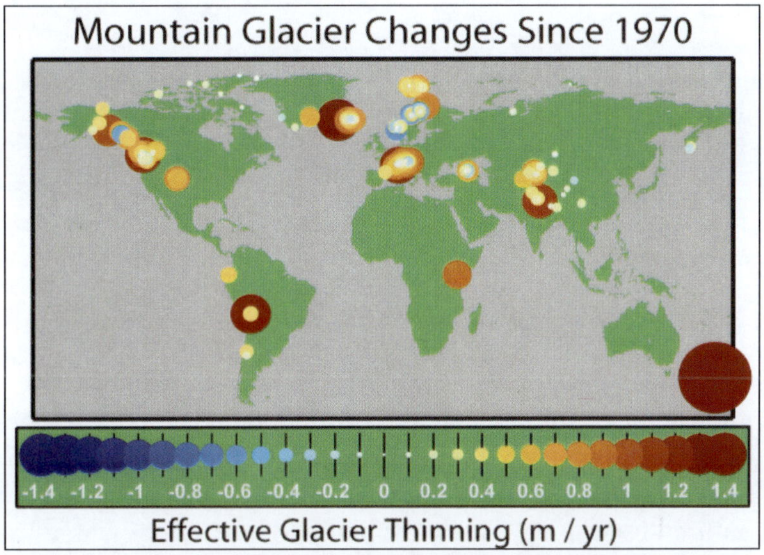

A map of the change in thickness of mountain glaciers since 1970. Thinning in orange and red, thickening in blue.

Warming temperatures lead to the melting of glaciers and ice sheets. IPCC found that, on average, mountain glaciers and snow cover had decreased in both the northern and southern hemispheres. This widespread decrease in glaciers and ice caps has contributed to observed sea level rise.

As stated above, the total volume of glaciers on Earth is declining sharply. Glaciers have been retreating worldwide for at least the last century; the rate of retreat has increased in the past decade. Only a few glaciers are actually advancing (in locations that were well below freezing, and where increased precipitation has outpaced melting). The progressive disappearance of glaciers has implications not only for a rising global sea level, but also for water supplies in certain regions of Asia and South America.

With very high or high confidence, IPCC made a number of projections related to future changes in glaciers:

- Mountainous areas in Europe will face glacier retreat.

- In Latin America, changes in precipitation patterns and the disappearance of glaciers will significantly affect water availability for human consumption, agriculture, and energy production.
- In Polar regions, there will be reductions in glacier extent and the thickness of glaciers.

In historic times, glaciers grew during a cool period from about 1550 to 1850 known as the Little Ice Age. Subsequently, until about 1940, glaciers around the world retreated as the climate warmed. Glacier retreat declined and reversed in many cases from 1950 to 1980 as a slight global cooling occurred. Since 1980, glacier retreat has become increasingly rapid and ubiquitous, and has threatened the existence of many of the glaciers of the world. This process has increased markedly since 1995.

Excluding the ice caps and ice sheets of the Arctic and Antarctic, the total surface area of glaciers worldwide has decreased by 50% since the end of the 19th century. Currently glacier retreat rates and mass balance losses have been increasing in the Andes, Alps, Pyrenees, Himalayas, Rocky Mountains and North Cascades.

The loss of glaciers not only directly causes landslides, flash floods and glacial lake overflow, but also increases annual variation in water flows in rivers. Glacier runoff declines in the summer as glaciers decrease in size, this decline is already observable in several regions. Glaciers retain water on mountains in high precipitation years, since the snow cover accumulating on glaciers protects the ice from melting. In warmer and drier years, glaciers offset the lower precipitation amounts with a higher meltwater input.

Some world regions, such as the French Alps, already show signs of an increase in landslide frequency.

Of particular importance are the Hindu Kush and Himalayan glacial melts that comprise the principal dry-season water source of many of the major rivers of the Central, South, East and Southeast Asian mainland. Increased melting would cause greater flow for several decades, after which "some areas of the most populated regions on Earth are likely to 'run out of water'" as source glaciers are depleted. The Tibetan Plateau contains the world's third-largest store of ice. Temperatures there are rising four times faster than in the rest of China, and glacial retreat is at a high speed compared to elsewhere in the world.

The recession of mountain glaciers, notably in Western North America, Franz-Josef Land, Asia, the Alps, the Pyrenees, Indonesia and Africa, and tropical and sub-tropical regions of South America, has been used to provide qualitative support to the rise in global temperatures since the late 19th century. Many glaciers are being lost to melting further raising concerns about future local water resources in these glaciated areas. In Western North America the 47 North Cascade glaciers observed all are retreating.

Despite their proximity and importance to human populations, the mountain and valley glaciers of temperate latitudes amount to a small fraction of glacial ice on the earth. About 99% is in the great ice sheets of polar and subpolar Antarctica and Greenland. These continuous continental-scale ice sheets, 3 kilometres (1.9 mi) or more in thickness, cap the polar and subpolar land masses. Like rivers flowing from an enormous lake, numerous outlet glaciers transport ice from the margins of the ice sheet to the ocean.

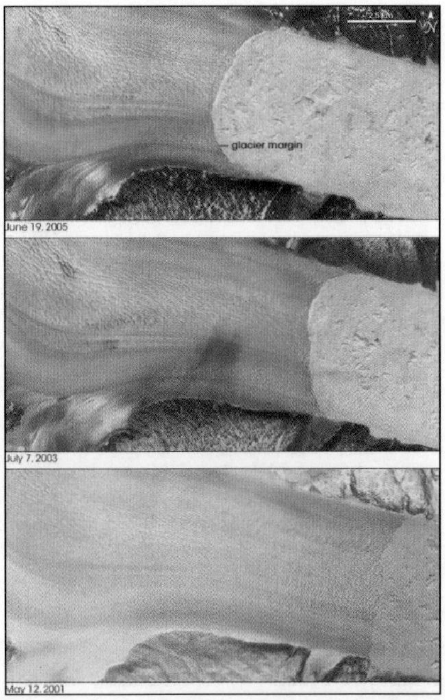

Retreat of the Helheim Glacier, Greenland.

Glacier retreat has been observed in these outlet glaciers, resulting in an increase of the ice flow rate. In Greenland the period since the year 2000 has brought retreat to several very large glaciers that had long been stable. Three glaciers that have been researched, Helheim, Jakobshavn Isbræ and Kangerdlugssuaq Glaciers, jointly drain more than 16% of the Greenland Ice Sheet. Satellite images and aerial photographs from the 1950s and 1970s show that the front of the glacier had remained in the same place for decades. But in 2001 it began retreating rapidly, retreating 7.2 km (4.5 mi) between 2001 and 2005. It has also accelerated from 20 m (66 ft)/day to 32 m (105 ft)/day. Jakobshavn Isbræ in western Greenland had been moving at speeds of over 24 m (79 ft)/day with a stable terminus since at least 1950. The glacier's ice tongue began to break apart in 2000, leading to almost complete disintegration in 2003, while the retreat rate increased to over 30 m (98 ft)/day.

Oceans

The role of the oceans in global warming is a complex one. The oceans serve as a sink for carbon dioxide, taking up much that would otherwise remain in the atmosphere, but increased levels of CO_2 have led to ocean acidification. Furthermore, as the temperature of the oceans increases, they become less able to absorb excess CO_2. Global warming is projected to have a number of effects on the oceans. Ongoing effects include rising sea levels due to thermal expansion and melting of glaciers and ice sheets, and warming of the ocean surface, leading to increased temperature stratification. Other possible effects include large-scale changes in ocean circulation.

Sea Level Rise

IPCC reported that since 1961, global average sea level had risen at an average rate of 1.8 (1.3 to 2.3) mm/yr. Between 1993 and 2003, the rate increased above the previous period to 3.1 (2.4 to 3.8) mm/yr. IPCC were uncertain whether the increase in rate from 1993 to 2003 was due to natural variations in sea level over the time period, or whether it reflected an increase in the underlying long-term trend.

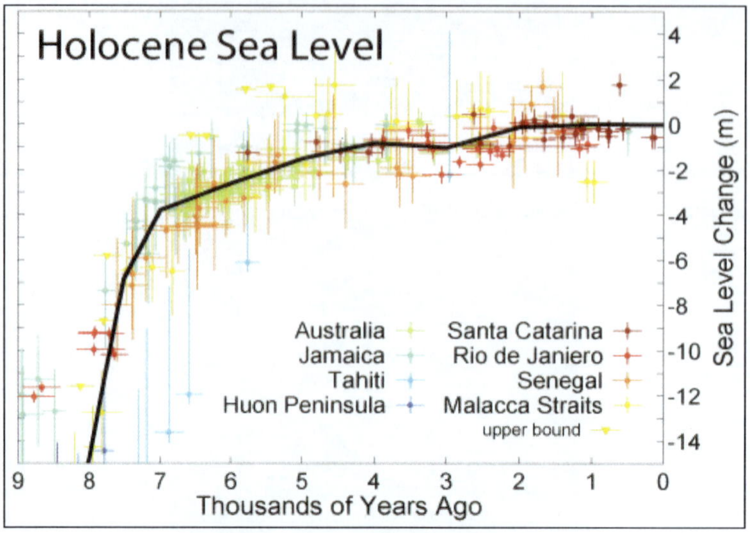

Sea level rise during the Holocene.

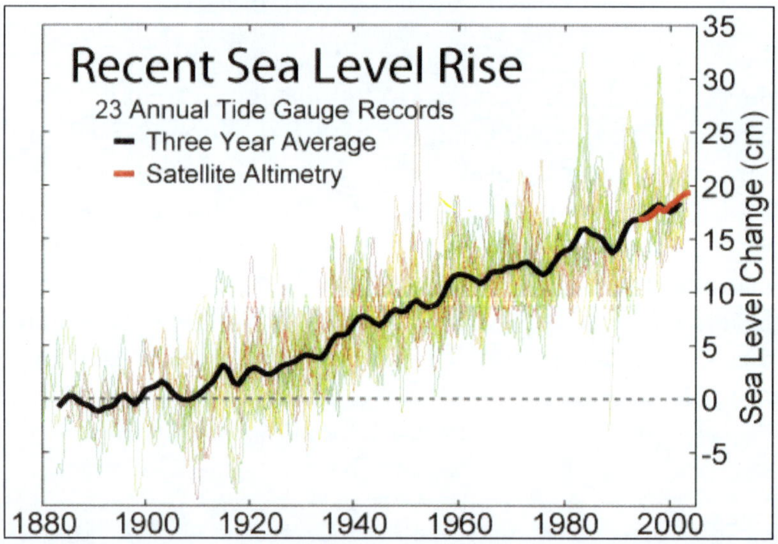

Sea level has been rising 0.2 cm/year, based on measurements of sea level rise from 23 long tide gauge records in geologically stable environments.

IPCC projected sea level rise to the end of the 21st century using the SRES emission scenarios. Across the six SRES marker scenarios, sea level was projected to rise by 18 to 59 cm (7.1 to 23.2 inches). This projection was for the time period 2090–2099, with the increase in level relative to average sea levels over the 1980–1999 period. Due to a lack of scientific understanding, this sea level rise estimate does not include all of the possible contributions of ice sheets.

With increasing average global temperature, the water in the oceans expands in volume, and additional water enters them which had previously been locked up on land in glaciers and ice sheets. The Greenland and the Antarctic ice sheets are major ice masses, and at least the former of which may suffer irreversible decline. For most glaciers worldwide, an average volume loss of 60% until 2050 is predicted. Meanwhile, the estimated total ice melting rate over Greenland is 239 ± 23 cubic kilometres (57.3 ± 5.5 cu mi) per year, mostly from East Greenland. The Antarctic ice sheet, however, is expected to grow during the 21st century because of increased precipitation. Under the IPCC Special Report on Emission Scenario (SRES) A1B, by the mid-2090s global sea level will reach 0.22 to 0.44 m (8.7 to 17.3 in) above 1990 levels, and is currently rising at about 4 mm (0.16 in) per year. Since 1900, the sea level has risen at an average of 1.7 mm (0.067 in) per year; since 1993, satellite altimetry from TOPEX/Poseidon indicates a rate of about 3 mm (0.12 in) per year.

The sea level has risen more than 120 metres (390 ft) since the Last Glacial Maximum about 20,000 years ago. The bulk of that occurred before 7000 years ago. Global temperature declined after the Holocene Climatic Optimum, causing a sea level lowering of 0.7 ± 0.1 m (27.6 ± 3.9 in) between 4000 and 2500 years before present. From 3000 years ago to the start of the 19th century, sea level was almost constant, with only minor fluctuations. However, the Medieval Warm Period may have caused some sea level rise; evidence has been found in the Pacific Ocean for a rise to perhaps 0.9 m (2 ft 11 in) above present level in 700 BP.

Sea level rise due to the collapse of an ice sheet would be distributed nonuniformly across the globe. The loss of mass in the region around the ice sheet would decrease the gravitational potential there, reducing the amount of local sea level rise or even causing local sea level fall. The loss of the localized mass would also change the moment of inertia of the Earth, as flow in the Earth's mantle will require 10–15 thousand years to make up the mass deficit. This change in the moment of inertia results in true polar wander, in which the Earth's rotational axis remains fixed with respect to the sun, but the rigid sphere of the Earth rotates with respect to it. This changes the location of the equatorial bulge of the Earth and further affects the geoid, or global potential field. A 2009 study of the effects of collapse of the West Antarctic Ice Sheet shows the result of both of these effects. Instead of a global 5-meter sea level rise, western Antarctica would experience approximately 25 centimeters of sea level fall, while the United States, parts of Canada, and the Indian Ocean, would experience up to 6.5 meters of sea level rise.

A paper published in 2008 by a group of researchers at the University of Wisconsin led by Anders Carlson used the deglaciation of North America at 9000 years before present as an analogue to predict sea level rise of 1.3 meters in the next century, which is also much higher than the IPCC projections. However, models of glacial flow in the smaller present-day ice sheets show that a probable maximum value for sea level rise in the next century is 80 centimeters, based on limitations on how quickly ice can flow below the equilibrium line altitude and to the sea.

Temperature Rise and Ocean Heat Content

Time series of seasonal (red dots) and annual average (black line) global upper ocean heat content for the 0-700m layer between 1955 and 2008. The graph shows that ocean heat content has increased over this time period.

From 1961 to 2003, the global ocean temperature has risen by 0.10 °C from the surface to a depth of 700 m. There is variability both year-to-year and over longer time scales, with global ocean heat content observations showing high rates of warming for 1991 to 2003, but some cooling from 2003 to 2007. Nevertheless, there is a strong trend during the period of reliable measurements. Increasing heat content in the ocean is also consistent with sea level rise, which is occurring mostly as a result of thermal expansion of the ocean water as it warms.

The temperature of the Antarctic Southern Ocean rose by 0.17 °C (0.31 °F) between the 1950s and the 1980s, nearly twice the rate for the world's oceans as a whole. As well as having effects on ecosystems (e.g. by melting sea ice, affecting algae that grow on its underside), warming reduces the ocean's ability to absorb CO_2.

Acidification

Ocean acidification is an effect of rising concentrations of CO_2 in the atmosphere, and is not a direct consequence of global warming. The oceans soak up much of the CO_2 produced by living organisms, either as dissolved gas, or in the skeletons of tiny marine creatures that fall to the bottom to become chalk or limestone. Oceans currently absorb about one tonne of CO_2 per person per year. It is estimated that the oceans have absorbed around half of all CO_2 generated by human activities since 1800.

In water, CO_2 becomes a weak carbonic acid, and the increase in the greenhouse gas since the Industrial Revolution has already lowered the average pH (the laboratory measure of acidity) of seawater by 0.1 units, to 8.2. Predicted emissions could lower the pH by a further 0.5 by 2100, to a level probably not seen for hundreds of millennia and, critically, at a rate of change probably 100 times greater than at any time over this period.

There are concerns that increasing acidification could have a particularly detrimental effect on corals (16% of the world's coral reefs have died from bleaching caused by warm water in 1998, which coincidentally was, at the time, the warmest year ever recorded) and other marine organisms with calcium carbonate shells.

In November 2009 an report in *Science* by scientists at Canada's Department of Fisheries and Oceans reported they had found very low levels of the building blocks for the calcium chloride that forms plankton shells in the Beaufort Sea. Fiona McLaughlin, one of the DFO authors, asserted that the increasing acidification of the Arctic Ocean was close to the point it would start dissolving the walls of existing plankton: "Arctic ecosystem may be risk. In actual fact, they'll dissolve the shells." Because cold water absorbs CO_2 more readily than warmer water the acidification is more severe in the polar regions. McLaughlin predicted the acidified water would travel to the North Atlantic within the next ten years.

Shutdown of Thermohaline Circulation

There is some speculation that global warming could, via a shutdown or slowdown of the thermohaline circulation, trigger localized cooling in the North Atlantic and lead to cooling, or lesser warming, in that region. This would affect in particular areas like Scandinavia and Britain that are warmed by the North Atlantic drift.

The chances of this near-term collapse of the circulation are unclear; there is some evidence for the short-term stability of the Gulf Stream and possible weakening of the North Atlantic drift. However, the degree of weakening, and whether it will be sufficient to shut down the circulation, is under debate. As yet, no cooling has been found in northern Europe or nearby seas. Lenton found that "simulations clearly pass a THC tipping point this century".

IPCC concluded that a slowing of the Meridional Overturning Circulation would very likely occur this century. Due to global warming, temperatures across the Atlantic and Europe were still projected to increase.

Oxygen Depletion

The amount of oxygen dissolved in the oceans may decline, with adverse consequences for ocean life.

Sulfur Aerosols

Sulfur aerosols, especially stratospheric sulfur aerosols have a significant effect on climate. One source of such aerosols is the sulfur cycle, where plankton release gases such as DMS which eventually becomes oxidised to sulfur dioxide in the atmosphere. Disruption to the oceans as a result of ocean acidification or disruptions to the thermohaline circulation may result in disruption of the sulfur cycle, thus reducing its cooling effect on the planet through the creation of stratospheric sulfur aerosols.

Geology

Volcanoes

The retreat of glaciers and ice caps can cause increased volcanism. Reduction in ice cover reduces the confining pressure exerted on the volcano, increasing deviatoric stresses and potentially causing the volcano to erupt. This reduction of pressure can also cause decompression melting of material in the mantle, resulting in the generation of more magma. Researchers in Iceland have shown that the rate of volcanic rock production there following deglaciation (10,000 to 4500 years before present) was 20–30 times greater than that observed after 2900 years before present. While the original study addresses the first reason for increased volcanism (reduced confining pressure), scientists have more recently shown that these lavas have unusually high trace element concentrations, indicative of increased melting in the mantle. This work in Iceland has been corroborated by a study in California, in which scientists found a strong correlation between volcanism and periods of global deglaciation. The effects of current sea level rise could include increased crustal stress at the base of coastal volcanoes from a rise in the volcano's water table (and the associated saltwater intrusion), while the mass from extra water could activate dormant seismic faults around volcanoes. In addition, the wide-scale displacement of water from melting in places such as West Antarctica is likely to slightly alter the Earth's rotational period and may shift its axial tilt on the scale of hundreds of metres, inducing further crustal stress changes.

Current melting of ice is predicted to increase the size and frequency of volcanic eruptions. In particular, lateral collapse events at stratovolcanoes are likely to increase, and there are potential positive feedbacks between the removal of ice and magmatism.

Earthquakes

A numerical modeling study has demonstrated that seismicity increases during unloading, such as that due to the removal of ice.

CLIMATE CHANGE AND AGRICULTURE

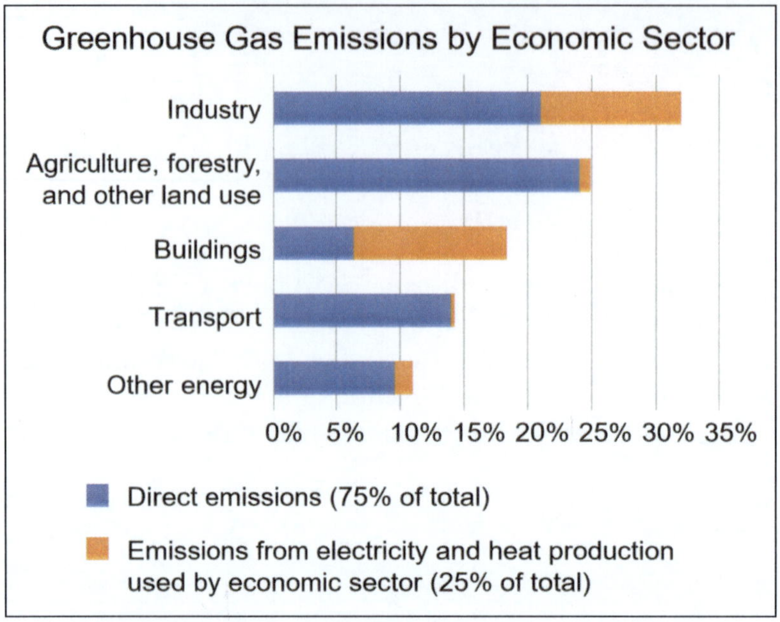

Human greenhouse gas emissions by sector, in the year 2010.

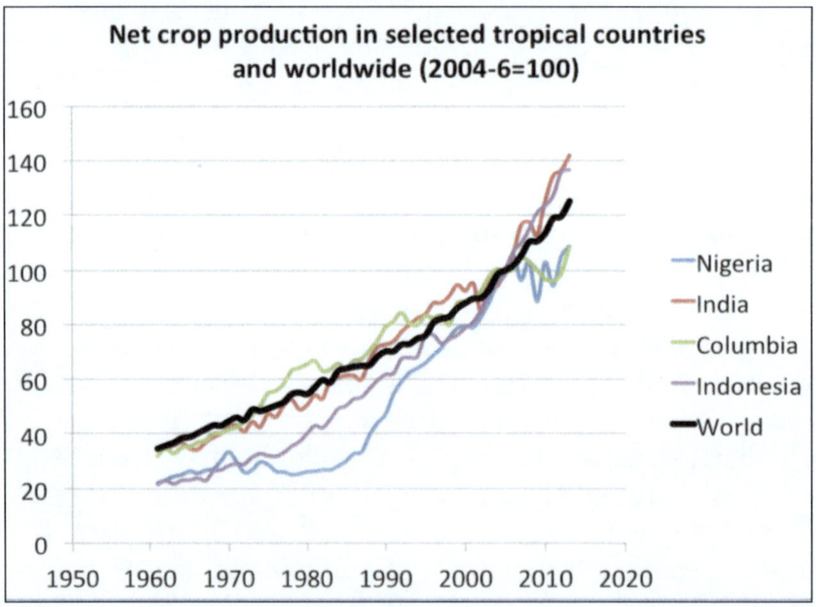

Graph of net crop production worldwide and in selected tropical countries. Raw data from the United Nations.

Climate change and agriculture are interrelated processes, both of which take place on a global scale. Climate change affects agriculture in a number of ways, including through changes in average temperatures, rainfall, and climate extremes (e.g., heat waves); changes in pests and diseases; changes in atmospheric carbon dioxide and ground-level ozone concentrations; changes in the nutritional quality of some foods; and changes in sea level.

Climate change is already affecting agriculture, with effects unevenly distributed across the world. Future climate change will likely negatively affect crop production in low latitude countries, while effects in northern latitudes may be positive or negative. Climate change will probably increase the risk of food insecurity for some vulnerable groups, such as the poor. Animal agriculture is also responsible for CO_2 greenhouse gas production and a percentage of the world's methane, and future land infertility, and the displacement of local species.

Agriculture contributes to climate change both by anthropogenic emissions of greenhouse gases and by the conversion of non-agricultural land such as forests into agricultural land. Agriculture, forestry and land-use change contributed around 20 to 25% of global annual emissions in 2010.

A range of policies can reduce the risk of negative climate change impacts on agriculture and greenhouse gas emissions from the agriculture sector.

Impact of Climate Change on Agriculture

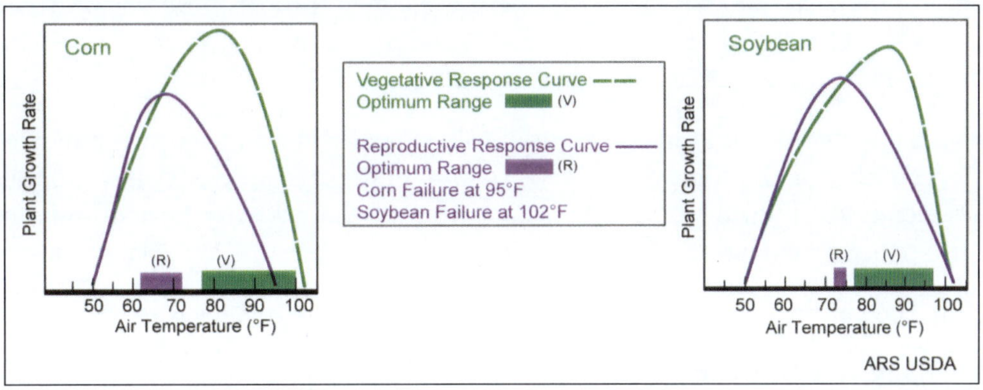

For each plant variety, there is an optimal temperature for vegetative growth, with growth dropping off as temperatures increase or decrease. Similarly, there is a range of temperatures at which a plant will produce seed. Outside of this range, the plant will not reproduce. As the graphs show, corn will fail to reproduce at temperatures above 95 °F (35 °C) and soybean above 102 °F (38.8 °C).

Despite technological advances, such as improved varieties, genetically modified organisms, and irrigation systems, weather is still a key factor in agricultural productivity, as well as soil properties and natural communities. The effect of climate on agriculture is related to variabilities in local climates rather than in global climate patterns. The Earth's average surface temperature has increased by 1.5 °F (0.83 °C) since 1880. Consequently, in making an assessment agronomists must consider each local area.

On the other hand, agricultural trade has grown in recent years, and now provides significant amounts of food, on a national level to major importing countries, as well as comfortable income to exporting ones. The international aspect of trade and security in terms of food implies the need to also consider the effects of climate change on a global scale.

Due to climate change, "southern Africa could lose more than 30% of its main crop, maize, by 2030. In South Asia losses of many regional staples, such as rice, millet and maize could top 10%".

The Intergovernmental Panel on Climate Change (IPCC) has produced several reports that have assessed the scientific literature on climate change. The IPCC Third Assessment Report, published in 2001, concluded that the poorest countries would be hardest hit, with reductions in crop yields in most tropical and sub-tropical regions due to decreased water availability, and new or changed insect pest incidence. In Africa and Latin America many rainfed crops are near their maximum temperature tolerance, so that yields are likely to fall sharply for even small climate changes; falls in agricultural productivity of up to 30% over the 21st century are projected. Marine life and the fishing industry will also be severely affected in some places.

In the report published in 2014 the Intergovernmental Panel on Climate Change claimed that the world may reach "a threshold of global warming beyond which current agricultural practices can no longer support large human civilizations". By the middle of the 21st century. In 2019 it published reports in which it says that millions already suffer from food insecurity due to climate change and predicted decline in global crop production of 2% - 6% by decade.

Climate change induced by increasing greenhouse gases is likely to affect crops differently from region to region. For example, average crop yield is expected to drop down to 50% in Pakistan according to the Met Office scenario whereas corn production in Europe is expected to grow up to 25% in optimum hydrologic conditions.

More favourable effects on yield tend to depend to a large extent on realization of the potentially beneficial effects of carbon dioxide on crop growth and increase of efficiency in water use. Decrease in potential yields is likely to be caused by shortening of the growing period, decrease in water availability and poor vernalization.

In the long run, the climatic change could affect agriculture in several ways :

- Productivity, in terms of quantity and quality of crops.
- Agricultural practices, through changes of water use (irrigation) and agricultural inputs such as herbicides, insecticides and fertilizers.
- Environmental effects, in particular in relation of frequency and intensity of soil drainage (leading to nitrogen leaching), soil erosion, reduction of crop diversity.
- Rural space, through the loss and gain of cultivated lands, land speculation, land renunciation, and hydraulic amenities.
- Adaptation, organisms may become more or less competitive, as well as humans may develop urgency to develop more competitive organisms, such as flood resistant or salt resistant varieties of rice.

They are large uncertainties to uncover, particularly because there is lack of information on many specific local regions, and include the uncertainties on magnitude of climate change, the effects of technological changes on productivity, global food demands, and the numerous possibilities of adaptation.

Most agronomists believe that agricultural production will be mostly affected by the severity and pace of climate change, not so much by gradual trends in climate. If change is gradual, there may be enough time for biota adjustment. Rapid climate change, however, could harm agriculture in many countries, especially those that are already suffering from rather poor soil and climate conditions, because there is less time for optimum natural selection and adaption.

But much remains unknown about exactly how climate change may affect farming and food security, in part because the role of farmer behaviour is poorly captured by crop-climate models. For instance, Evan Fraser, a geographer at the University of Guelph in Ontario Canada, has conducted a number of studies that show that the socio-economic context of farming may play a huge role in determining whether a drought has a major, or an insignificant impact on crop production. In some cases, it seems that even minor droughts have big impacts on food security (such as what happened in Ethiopia in the early 1980s where a minor drought triggered a massive famine), versus cases where even relatively large weather-related problems were adapted to without much hardship. Evan Fraser combines socio-economic models along with climatic models to identify "vulnerability hotspots" One such study has identified US maize (corn) production as particularly vulnerable to climate change because it is expected to be exposed to worse droughts, but it does not have the socio-economic conditions that suggest farmers will adapt to these changing conditions. Other studies rely instead on projections of key agro-meteorological or agro-climate indices, such as growing season length, plant heat stress, or start of field operations, identified by land management stakeholders and that provide useful information on mechanisms driving climate change impact on agriculture.

Pest Insects and Climate Change

Global warming could lead to an increase in pest insect populations, harming yields of staple crops like wheat, soybeans, and corn. While warmer temperatures create longer growing seasons, and faster growth rates for plants, it also increases the metabolic rate and number of breeding cycles of insect populations. Insects that previously had only two breeding cycles per year could gain an additional cycle if warm growing seasons extend, causing a population boom. Temperate places and higher latitudes are more likely to experience a dramatic change in insect populations.

The University of Illinois conducted studies to measure the effect of warmer temperatures on soybean plant growth and Japanese beetle populations. Warmer temperatures and elevated CO_2 levels were simulated for one field of soybeans, while the other was left as a control. These studies found that the soybeans with elevated CO_2 levels grew much faster and had higher yields, but attracted Japanese beetles at a significantly higher rate than the control field. The beetles in the field with increased CO_2 also laid more eggs on the soybean plants and had longer lifespans, indicating the possibility of a rapidly expanding population. DeLucia projected that if the project were to continue, the field with elevated CO_2 levels would eventually show lower yields than that of the control field.

The increased CO_2 levels deactivated three genes within the soybean plant that normally create chemical defenses against pest insects. One of these defenses is a protein that blocks digestion of the soy leaves in insects. Since this gene was deactivated, the beetles were able to digest a much higher amount of plant matter than the beetles in the control field. This led to the observed longer lifespans and higher egg-laying rates in the experimental field.

There are a few proposed solutions to the issue of expanding pest populations. One proposed solution is to increase the number of pesticides used on future crops. This has the benefit of being relatively cost effective and simple, but may be ineffective. Many pest insects have been building up an immunity to these pesticides. Another proposed solution is to utilize biological control agents. This includes things like planting rows of native vegetation in between rows of crops. This solution is beneficial in its overall environmental impact. Not only are more native plants getting planted, but pest insects are no longer building up an immunity to pesticides. However, planting additional native plants requires more room, which destroys additional acres of public land. The cost is also much higher than simply using pesticides.

Plant Diseases and Climate Change

Although research is limited, research has shown that climate change may alter the developmental stages of pathogens that can affect crops. The biggest consequence of climate change on the dispersal of pathogens is that the geographical distribution of hosts and pathogens could shift, which would result in more crop losses. This could affect competition and recovery from disturbances of plants. It has been predicted that the effect of climate change will add a level of complexity to figuring out how to maintain sustainable agriculture.

Observed Impacts

Effects of regional climate change on agriculture have been limited. Changes in crop phenology provide important evidence of the response to recent regional climate change. Phenology is the study of natural phenomena that recur periodically, and how these phenomena relate to climate and seasonal changes. A significant advance in phenology has been observed for agriculture and forestry in large parts of the Northern Hemisphere.

Droughts have been occurring more frequently because of global warming and they are expected to become more frequent and intense in Africa, southern Europe, the Middle East, most of the Americas, Australia, and Southeast Asia. Their impacts are aggravated because of increased water demand, population growth, urban expansion, and environmental protection efforts in many areas. Droughts result in crop failures and the loss of pasture grazing land for livestock.

Examples,

Banana farm at Chinawal village in Jalgaon district, India.

As of the decade starting in 2010, many hot countries have thriving agricultural sectors.

Jalgaon district, India, has an average temperature which ranges from 20.2 °C in December to 29.8 °C in May, and an average precipitation of 750 mm/year. It produces bananas at a rate that would make it the world's seventh-largest banana producer if it were a country.

During the period 1990-2012, Nigeria had an average temperature which ranged from a low of 24.9 °C in January to a high of 30.4 °C in April. According to the Food and Agriculture Organization of the United Nations (FAO), Nigeria is by far the world's largest producer of yams, producing over 38 million tonnes in 2012. The second through 8th largest yam producers were all nearby African countries, with the largest non-African producer, Papua New Guinea, producing less than 1% of Nigerian production.

In 2013, according to the FAO, Brazil and India were by far the world's leading producers of sugarcane, with a combined production of over 1 billion tonnes, or over half of worldwide production.

In the summer of 2018, heat waves probably linked to climate change cause much lower than average yield in many parts of the world, especially in Europe. Depending on conditions during August, more crop failures could rise global food prices. losses are compared to those of 1945, the worst harvest in memory. last year was the third time in four years that global wheat, rice and maize production failed to meet demand, forcing governments and food companies to release stocks from storage. India last week released 50% of its food stocks. Lester Brown, the head of Worldwatch, an independent research organisation, predicted that food prices will rise in the next few months.

According to the UN report "Climate Change and Land: an IPCC special report on climate change, desertification, land degradation, sustainable land management, food security, and greenhouse gas fluxes in terrestrial ecosystems", food prices will rise by 80% by 2050 and food shortages are likely to occur. Some authors also suggested that the food shortages will probably affect poorer parts of the world far more than richer ones.

To prevent hunger, instability, new waves of climate refugees, international help will be needed to countries who will miss the money to buy enough food and for also for stopping conflicts.

At the beginning of the 21 century, floods probably linked to climate change shortened the planting season in the Midwest region in United States, what cause damage to the agriculture sector. In May 2019 the floods reduced the projected Corn yield from 15 billion bushels to 14.2.

Projections of Impacts

As part of the IPCC's Fourth Assessment Report, Schneider projected the potential future effects of climate change on agriculture. With low to medium confidence, they concluded that for about a 1 to 3 °C global mean temperature increase there would be productivity decreases for some cereals in low latitudes, and productivity increases in high latitudes. In the IPCC Fourth Assessment Report, "low confidence" means that a particular finding has about a 2 out of 10 chance of being correct, based on expert judgement. "Medium confidence" has about a 5 out of 10 chance of being correct. Over the same time period, with medium confidence, global production potential was projected to:

- Increase up to around 3 °C,
- Very likely decrease above about 3 °C.

Most of the studies on global agriculture assessed by Schneider had not incorporated a number of critical factors, including changes in extreme events, or the spread of pests and diseases. Studies had also not considered the development of specific practices or technologies to aid adaptation to climate change.

The US National Research Council assessed the literature on the effects of climate change on crop yields. US NRC stressed the uncertainties in their projections of changes in crop yields.

Matthew Smith and Samuel Myers estimated that food crops could see a reduction of protein, iron and zinc content in common food crops of 3 to 17%. This is the projected result of food grown under the expected atmospheric carbon-dioxide levels of 2050. Using data from the UN Food and Agriculture Organization as well as other public sources, the authors analyzed 225 different staple foods, such as wheat, rice, maize, vegetables, roots and fruits.

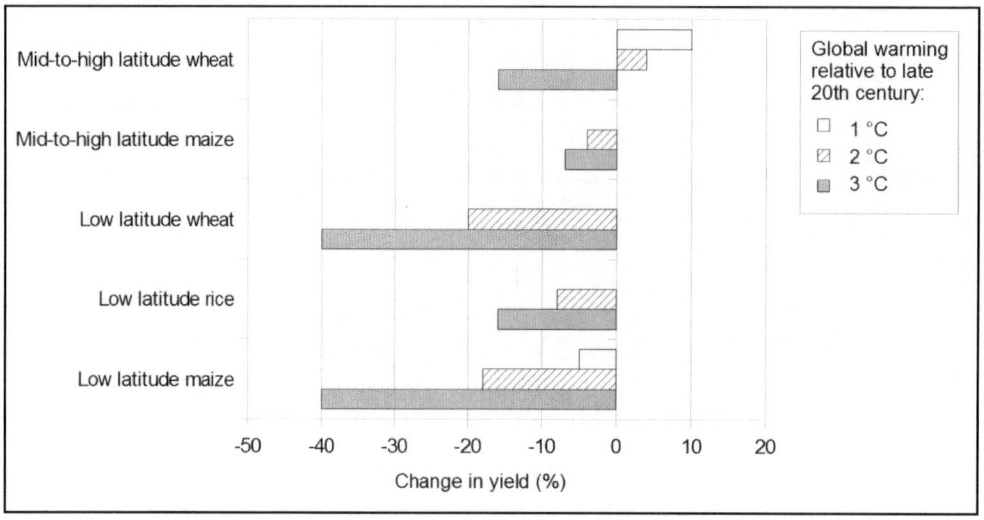

Projected changes in crop yields at different latitudes with global warming. This graph is based on several studies.

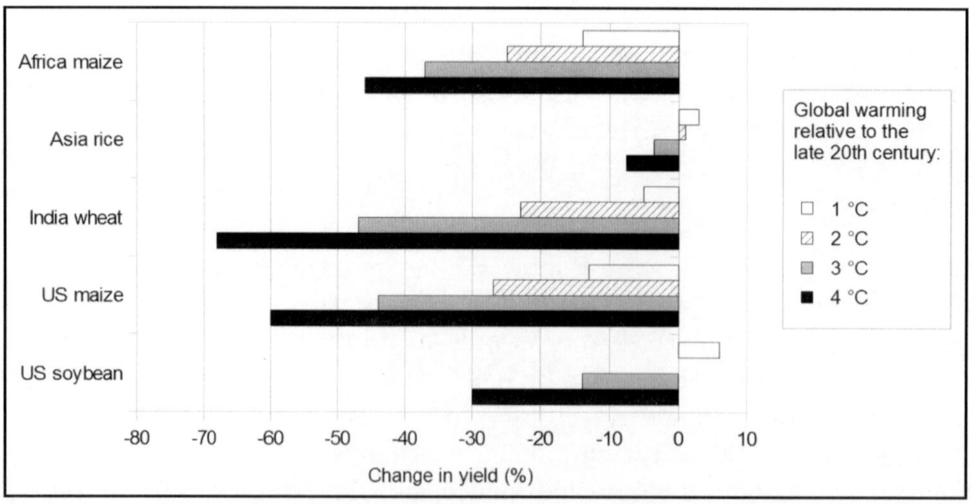

Projected changes in yields of selected crops with global warming. This graph is based on several studies.

Their central estimates of changes in crop yields are shown above. Actual changes in yields may be above or below these central estimates. US NRC) also provided an estimated the "likely" range of changes in yields. "Likely" means a greater than 67% chance of being correct, based on expert judgement. The likely ranges are summarized in the image descriptions of the two graphs.

Food Security

The IPCC Fourth Assessment Report also describes the impact of climate change on food security. Projections suggested that there could be large decreases in hunger globally by 2080, compared to the 2006 level. Reductions in hunger were driven by projected social and economic development. For reference, the Food and Agriculture Organization has estimated that in 2006, the number of people undernourished globally was 820 million. Three scenarios without climate change (SRES A1, B1, B2) projected 100-130 million undernourished by the year 2080, while another scenario without climate change (SRES A2) projected 770 million undernourished. Based on an expert assessment of all of the evidence, these projections were thought to have about a 5-in-10 chance of being correct.

The same set of greenhouse gas and socio-economic scenarios were also used in projections that included the effects of climate change. *Including* climate change, three scenarios (SRES A1, B1, B2) projected 100-380 million undernourished by the year 2080, while another scenario with climate change (SRES A2) projected 740-1,300 million undernourished. These projections were thought to have between a 2-in-10 and 5-in-10 chance of being correct.

Projections also suggested regional changes in the global distribution of hunger. By 2080, sub-Saharan Africa may overtake Asia as the world's most food-insecure region. This is mainly due to projected social and economic changes, rather than climate change.

In South America, a phenomenon known as the El Nino Oscillation Cycle, between floods and drought on the Pacific Coast has made as much as a 35% difference in Global yields of wheat and grain.

Looking at the four key components of food security we can see the impact climate change has had. Food "Access to food is largely a matter of household and individual-level income and of capabilities and rights". Access has been affected by the thousands of crops being destroyed, how communities are dealing with climate shocks and adapting to climate change. Prices on food will rise due to the shortage of food production due to conditions not being favourable for crop production. Utilization is affected by floods and drought where water resources are contaminated, and the changing temperatures create vicious stages and phases of disease. Availability is affected by the contamination of the crops, as there will be no food process for the products of these crops as a result. Stability is affected through price ranges and future prices as some food sources are becoming scarce due to climate change, so prices will rise.

Individual Studies

Cline looked at how climate change might affect agricultural productivity in the 2080s. His study assumes that no efforts are made to reduce anthropogenic greenhouse gas emissions, leading to global warming of 3.3 °C above the pre-industrial level. He concluded that global agricultural productivity could be negatively affected by climate change, with the worst effects in developing countries.

Lobell assessed how climate change might affect 12 food-insecure regions in 2030. The purpose of their analysis was to assess where adaptation measures to climate change should be prioritized. They found that without sufficient adaptation measures, South Asia and South Africa would likely suffer negative impacts on several crops which are important to large food insecure human populations.

Battisti and Naylor looked at how increased seasonal temperatures might affect agricultural productivity. Projections by the IPCC suggest that with climate change, high seasonal temperatures will become widespread, with the likelihood of extreme temperatures increasing through the second half of the 21st century. Battisti and Naylor concluded that such changes could have very serious effects on agriculture, particularly in the tropics. They suggest that major, near-term, investments in adaptation measures could reduce these risks.

"Climate change merely increases the urgency of reforming trade policies to ensure that global food security needs are met" said C. Bellmann, ICTSD Programmes Director. A 2009 ICTSD-IPC study by Jodie Keane suggests that climate change could cause farm output in sub-Saharan Africa to decrease by 12% by 2080 - although in some African countries this figure could be as much as 60%, with agricultural exports declining by up to one fifth in others. Adapting to climate change could cost the agriculture sector $14bn globally a year, the study finds.

Regional Impacts

Africa

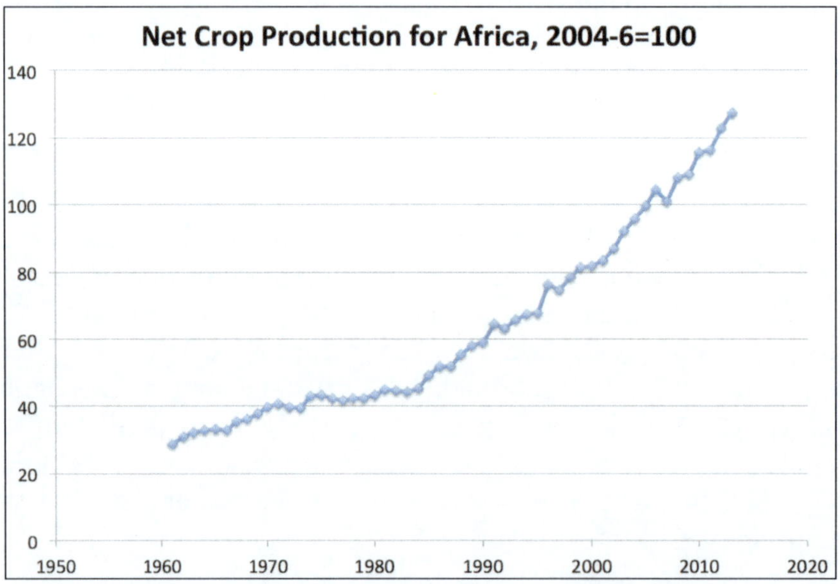

African crop production. Raw data from the United Nations.

Agriculture is a particularly important sector in Africa, contributing towards livelihoods and economies across the continent. On average, agriculture in Sub-Saharan Africa contributes 15% of the total GDP. Africa's geography makes it particularly vulnerable to climate change, and seventy per cent of the population rely on rain-fed agriculture for their livelihoods. Smallholder farms account for 80% of cultivated lands in sub-Saharan Africa. The Intergovernmental Panel on Climate Change (IPCC) projected that climate variability and change would severely compromise agricultural production and access to food. This projection was assigned "high confidence". Cropping systems, livestock and fisheries will be at greater risk of pest and diseases as a result of future climate change. Research program on Climate Change, Agriculture and Food Security (CCAFS) have identified that crop pests already account for approximately 1/6th of farm productivity losses. Similarly, climate change will accelerate the prevalence of pests and diseases and increase the occurrence of highly impactful events . The impacts of climate change on agricultural production in Africa will have serious implications for food security and livelihoods. Between 2014 and 2018, Africa had the highest levels of food insecurity in the world.

East Africa

In East Africa, climate change is anticipated to intensify the frequency and intensity of drought and flooding, which can have an adverse impact on the agricultural sector. Climate change will have varying effects on agricultural production in East Africa. Research from the International Food Policy Research Institute (IFPRI) suggest an increase in maize yields for most East Africa, but yield losses in parts of Ethiopia, Democratic Republic of Congo (DRC), Tanzania and northern Uganda. Projections of climate change are also anticipated to reduce the potential of the cultivated land to produce crops of high quantity and quality.

In Tanzania there is currently no clear signal in future climate projections for rainfall. However, there is a higher likelihood of intense future rainfall events. As of 2005, the net result was expected to be that 33% less maize—the country's staple crop—would be grown.

Southern Africa

Climate change will exacerbate the vulnerability of the Agricultural sector in most Southern African countries which are already limited by poor infrastructure and a lag in technological inputs and innovation . Maize accounts for nearly half of the cultivated land in Southern Africa, and under future climate change, yields could decrease by 30% Temperatures increases also encourage a wide spread of weeds and pests.

West Africa

Climate change will significantly affect agriculture in West Africa by increasing the variability in food production, access and availability. The region has already experienced a decrease in rainfall along the coasts of Nigeria, Sierra Leone, Guinea and Liberia. This has resulted in lower crop yield, causing farmers to seek new areas for cultivation. Staple crops such as maize, rice and sorghum will be impacted by low rainfall events with possible increase in food insecurity.

Central Africa

Higher rainfall intensity, prolonged dry spells and high temperatures are expected to negatively impact cassava, maize and bean production in Central Africa. Floods and erosion occurrence are expected to damage the already limited transportation infrastructure in the region leading to post harvest losses. Exportation of economic crops like coffee and cocoa are on the rise within the region but these crops are highly vulnerable to climate change. Conflicts and political instability have had an impact on agriculture contribution to the regional GDP and this impact will be exacerbated by climatic risks.

Asia

In East and Southeast Asia, IPCC projected that crop yields could increase up to 20% by the mid-21st century. In Central and South Asia, projections suggested that yields might decrease by up to 30%, over the same time period. These projections were assigned "medium confidence." Taken together, the risk of hunger was projected to remain very high in several developing countries.

More detailed analysis of rice yields by the International Rice Research Institute forecast 20% reduction in yields over the region per degree Celsius of temperature rise. Rice becomes sterile if exposed to temperatures above 35 degrees for more than one hour during flowering and consequently produces no grain.

A 2013 study by the International Crops Research Institute for the Semi-Arid Tropics (ICRISAT) aimed to find science-based, pro-poor approaches and techniques that would enable Asia's agricultural systems to cope with climate change, while benefitting poor and vulnerable farmers. The study's recommendations ranged from improving the use of climate information in local planning and strengthening weather-based agro-advisory services, to stimulating diversification of rural household incomes and providing incentives to farmers to adopt natural resource conservation measures to enhance forest cover, replenish groundwater and use renewable energy. A 2014 study found that warming had increased maize yields in the Heilongjiang region of China had increased by between 7 and 17% per decade as a result of rising temperatures.

Due to climate change, livestock production will be decreased in Bangladesh by diseases, scarcity of forage, heat stress and breeding strategies.

Australia and New Zealand

Hennessy assessed the literature for Australia and New Zealand. They concluded that without further adaptation to climate change, projected impacts would likely be substantial: By 2030, production from agriculture and forestry was projected to decline over much of southern and eastern Australia, and over parts of eastern New Zealand; In New Zealand, initial benefits were projected close to major rivers and in western and southern areas. Hennessy placed high confidence in these projections.

Europe

With high confidence, IPCC projected that in Southern Europe, climate change would reduce crop productivity. In Central and Eastern Europe, forest productivity was expected to decline. In Northern Europe, the initial effect of climate change was projected to increase crop yields. The 2019 European Environment Agency report "Climate change adaptation in the agricultural sector in Europe" again confirmed this. According to this 2019 report, projections indicate that yields of non-irrigated crops like wheat, corn and sugar beet would decrease in southern Europe by up to 50 % by 2050 (under a high-end emission scenario). This could result in a substantial decrease in farm income by that date. Also farmland values are projected to decrease in parts of southern

Europe by more than 80 % by 2100, which could result in land abandonment. The trade patterns are also said to be impacted, in turn affecting agricultural income. Also, increased food demand worldwide could exert pressure on food prices in the coming decades.

Latin America

The major agricultural products of Latin American regions include livestock and grains, such as maize, wheat, soybeans, and rice. Increased temperatures and altered hydrological cycles are predicted to translate to shorter growing seasons, overall reduced biomass production, and lower grain yields. Brazil, Mexico and Argentina alone contribute 70-90% of the total agricultural production in Latin America. In these and other dry regions, maize production is expected to decrease. A study summarizing a number of impact studies of climate change on agriculture in Latin America indicated that wheat is expected to decrease in Brazil, Argentina and Uruguay. Livestock, which is the main agricultural product for parts of Argentina, Uruguay, southern Brazil, Venezuela, and Colombia is likely to be reduced. Variability in the degree of production decrease among different regions of Latin America is likely. For example, one 2003 study that estimated future maize production in Latin America predicted that by 2055 maize in eastern Brazil will have moderate changes while Venezuela is expected to have drastic decreases.

Suggested potential adaptation strategies to mitigate the impacts of global warming on agriculture in Latin America include using plant breeding technologies and installing irrigation infrastructure.

North America

A number of studies have been produced which assess the impacts of climate change on agriculture in North America. The IPCC Fourth Assessment Report of agricultural impacts in the region cites 26 different studies. With high confidence, IPCC projected that over the first few decades of this century, moderate climate change would increase aggregate yields of rain-fed agriculture by 5–20%, but with important variability among regions. Major challenges were projected for crops that are near the warm end of their suitable range or which depend on highly utilized water resources.

Droughts are becoming more frequent and intense in arid and semiarid western North America as temperatures have been rising, advancing the timing and magnitude of spring snow melt floods and reducing river flow volume in summer. Direct effects of climate change include increased heat and water stress, altered crop phenology, and disrupted symbiotic interactions. These effects may be exacerbated by climate changes in river flow, and the combined effects are likely to reduce the abundance of native trees in favour of non-native herbaceous and drought-tolerant competitors, reduce the habitat quality for many native animals, and slow litter decomposition and nutrient cycling. Climate change effects on human water demand and irrigation may intensify these effects.

The US Global Change Research Program assessed the literature on the impacts of climate change on agriculture in the United States, finding that many crops will benefit from increased atmospheric CO_2 concentrations and low levels of warming, but that higher levels of warming will negatively affect growth and yields; that extreme weather events will likely reduce crop yields; that weeds, diseases and insect pests will benefit from warming, and will require additional pest and weed control; and that increasing CO_2 concentrations will reduce the land's ability to supply adequate livestock feed, while increased heat, disease, and weather extremes will likely reduce livestock productivity.

Polar Regions

Anisimov assessed the literature for the polar region (Arctic and Antarctica). With medium confidence, they concluded that the benefits of a less severe climate were dependent on local conditions. One of these benefits was judged to be increased agricultural and forestry opportunities.

A study reported on how climate change had affected agriculture in Iceland. Rising temperatures had made the widespread sowing of barley possible, which had been untenable twenty years ago. Some of the warming was due to a local (possibly temporary) effect via ocean currents from the Caribbean, which had also affected fish stocks.

Small Islands

In a literature assessment, Mimura concluded that on small islands, subsistence and commercial agriculture would very likely be adversely affected by climate change. This projection was assigned "high confidence".

Climate Change Impacts Affecting Poverty Alleviation

Researchers at the Overseas Development Institute (ODI) have investigated the potential impacts climate change could have on agriculture, and how this would affect attempts at alleviating poverty in the developing world. They argued that the effects from moderate climate change are likely to be mixed for developing countries. However, the vulnerability of the poor in developing countries to short term impacts from climate change, notably the increased frequency and severity of adverse weather events is likely to have a negative impact. This, they say, should be taken into account when defining agricultural policy.

Crop Development Models

Models for climate behavior are frequently inconclusive. In order to further study effects of global warming on agriculture, other types of models, such as *crop development models, yield prediction*, quantities of *water or fertilizer consumed*, can be used. Such models condense the knowledge accumulated of the climate, soil, and effects observed of the results of various agricultural practices. They thus could make it possible to test strategies of adaptation to modifications of the environment.

Because these models are necessarily simplifying natural conditions (often based on the assumption that weeds, disease and insect pests are controlled), it is not clear whether the results they give will have an *in-field* reality. However, some results are partly validated with an increasing number of experimental results.

Other models, such as *insect and disease development* models based on climate projections are also used (for example simulation of aphid reproduction or septoria (cereal fungal disease) development).

Scenarios are used in order to estimate climate changes effects on crop development and yield. Each scenario is defined as a set of meteorological variables, based on generally accepted projections. For example, many models are running simulations based on doubled carbon dioxide projections, temperatures raise ranging from 1 °C up to 5 °C, and with rainfall levels an increase or decrease of 20%. Other parameters may include humidity, wind, and solar activity. Scenarios of crop models are testing farm-level adaptation, such as sowing date shift, climate adapted species (vernalisation need, heat and cold resistance), irrigation and fertilizer adaptation, resistance to disease. Most developed models are about wheat, maize, rice and soybean.

Temperature Potential Effect on Growing Period

Duration of crop growth cycles are above all, related to temperature. An increase in temperature will speed up development. In the case of an annual crop, the duration between sowing and harvesting will shorten (for example, the duration in order to harvest corn could shorten between one and four weeks). The shortening of such a cycle could have an adverse effect on productivity because senescence would occur sooner.

Effect of Elevated Carbon Dioxide on Crops

Elevated atmospheric carbon dioxide affects plants in a variety of ways. Elevated CO_2 increases crop yields and growth through an increase in photosynthetic rate, and it also decreases water loss as a result of stomatal closing .It limits the vaporization of water reaching the stem of the plant. "Crassulacean Acid Metabolism" oxygen is all along the layer of the leaves for each plant leaves taking in $CO_2$and release O_2. The growth response is greatest in C_3 plants, C_4 plants,are also enhanced but to a lesser extent, and CAM Plants are the least enhanced species. The stoma in these "CAM plant" stores remain shut all day to reduce exposure. rapidly rising levels of carbon dioxide in the atmosphere affect plants' absorption of nitrogen, which is the nutrient that restricts crop growth in most terrestrial ecosystems. Today's concentration of 400 ppm plants are relatively starved for nutrition. The optimum level of CO_2 for plant growth is about 5 times higher. Increased mass of CO_2 increases photosynthesis, this CO_2 potentially stunts the growth of the plant. It limit's the reduction that crops lose through transpiration.

Increase in global temperatures will cause an increase in evaporation rates and annual evaporation levels. Increased evaporation will lead to an increase in storms in some areas, while leading to accelerated drying of other areas. These storm impacted areas will likely experience increased levels of precipitation and increased flood risks, while areas outside of the storm track will experience less precipitation and increased risk of droughts. Water stress effects plant development and quality in a variety of ways first off drought can cause poor germination and impaired seedling development in plants. At the same time plant growth relies on cellular division, cell enlargement, and differentiation. Drought stress impairs mitosis and cell elongation via loss of turgor pressure which results in poor growth. Development of leaves is also dependent upon turgor pressure, concentration of nutrients, and carbon assimilates all of which are reduced by drought conditions, thus drought stress lead to a decrease in leaf size and number. Plant height, biomass, leaf size and stem girth has been shown to decrease in Maize under water limiting conditions. Crop yield is also negatively effected by drought stress, the reduction in crop yield results from a decrease in photosynthetic rate, changes in leaf development, and altered allocation of resources all due to drought stress. Crop plants exposed to drought stress suffer from reductions in leaf water potential and transpiration rate, however water-use efficiency has been shown to increase in some crop plants such as wheat while decreasing in others such as potatoes. Plants need water for the uptake of nutrients from the soil, and for the transport of nutrients throughout the plant, drought conditions limit these functions leading to stunted growth. Drought stress also causes a decrease in photosynthetic activity in plants due to the reduction of photosynthetic tissues, stomatal closure, and reduced performance of photosynthetic machinery. This reduction in photosynthetic activity contributes to the reduction in plant growth and yields. Another factor influencing reduced plant growth and yields include the allocation of resources; following drought stress plants will allocate more resources to roots to aid in water uptake increasing root growth and reducing the growth of other plant parts while decreasing yields.

Effect on Quality

According to the IPCC's TAR, "The importance of climate change impacts on grain and forage quality emerges from new research. For rice, the amylose content of the grain—a major determinant of cooking quality—is increased under elevated CO_2". Cooked rice grain from plants grown in high-CO_2 environments would be firmer than that from today's plants. However, concentrations of iron and zinc, which are important for human nutrition, would be lower. Moreover, the protein content of the grain decreases under combined increases of temperature and CO_2. Studies using FACE have shown that increases in CO_2 lead to decreased concentrations of micronutrients in crop plants, including decreased B vitamins in rice. This may have knock-on effects on other parts of ecosystems as herbivores will need to eat more food to gain the same amount of protein.

Studies have shown that higher CO_2 levels lead to reduced plant uptake of nitrogen (and a smaller number showing the same for trace elements such as zinc) resulting in crops with lower nutritional value. This would primarily impact on populations in poorer countries less able to compensate by eating more food, more varied diets, or possibly taking supplements.

Reduced nitrogen content in grazing plants has also been shown to reduce animal productivity in sheep, which depend on microbes in their gut to digest plants, which in turn depend on nitrogen intake. Because of the lack of water available to crops in warmer countries they struggle to survive as they suffer from dehydration, taking into account the increasing demand for water outside of agriculture as well as other agricultural demands.

Effect of Hail

In North America, fewer hail days will occur overall due to climate change, but storms with larger hail might become more common (including hail that is larger than 1.6 inch). Hail that is larger than 1.6 inch can quite easily break (glass) greenhouses.

Agricultural Surfaces and Climate Changes

Climate change may increase the amount of arable land in high-latitude region by reduction of the amount of frozen lands. A 2005 study reports that temperature in Siberia has increased three degree Celsius in average since 1960 (much more than the rest of the world). However, reports about the impact of global warming on Russian agriculture indicate conflicting probable effects: while they expect a northward extension of farmable lands, they also warn of possible productivity losses and increased risk of drought.

Sea levels are expected to get up to one meter higher by 2100, though this projection is disputed. A rise in the sea level would result in an agricultural land loss, in particular in areas such as South East Asia. Erosion, submergence of shorelines, salinity of the water table due to the increased sea levels, could mainly affect agriculture through inundation of low-lying lands.

Low-lying areas such as Bangladesh, India and Vietnam will experience major loss of rice crop if sea levels rise as expected by the end of the century. Vietnam for example relies heavily on its southern tip, where the Mekong Delta lies, for rice planting. Any rise in sea level of no more than a meter will drown several km^2 of rice paddies, rendering Vietnam incapable of producing its main staple and export of rice.

Erosion and Fertility

The warmer atmospheric temperatures observed over the past decades are expected to lead to a more vigorous hydrological cycle, including more extreme rainfall events. Erosion and soil degradation is more likely to occur. Soil fertility would also be affected by global warming. Increased erosion in agricultural landscapes from anthropogenic factors can occur with losses of up to 22% of soil carbon in 50 years. However, because the ratio of soil organic carbon to nitrogen is mediated by soil biology such that it maintains a narrow range, a doubling of soil organic carbon is likely to imply a doubling in the storage of nitrogen in soils as organic nitrogen, thus providing higher available nutrient levels for plants, supporting higher yield potential. The demand for imported fertilizer nitrogen could decrease, and provide the opportunity for changing costly fertilisation strategies.

Due to the extremes of climate that would result, the increase in precipitations would probably result in greater risks of erosion, whilst at the same time providing soil with better hydration, according to the intensity of the rain. The possible evolution of the organic matter in the soil is a highly contested issue: while the increase in the temperature would induce a greater rate in the production of minerals, lessening the soil organic matter content, the atmospheric CO_2 concentration would tend to increase it.

Potential Effects of Global Climate Change on Pests, Diseases and Weeds

A very important point to consider is that weeds would undergo the same acceleration of cycle as cultivated crops, and would also benefit from carbonaceous fertilization. Since most weeds are C3 plants, they are likely to compete even more than now against C4 crops such as corn. However, on the other hand, some results make it possible to think that weedkillers could increase in effectiveness with the temperature increase.

Global warming would cause an increase in rainfall in some areas, which would lead to an increase of atmospheric humidity and the duration of the wet seasons. Combined with higher temperatures, these could favour the development of fungal diseases. Similarly, because of higher temperatures and humidity, there could be an increased pressure from insects and disease vectors.

Glacier Retreat and Disappearance

The continued retreat of glaciers will have a number of different quantitative impacts. In the areas that are heavily dependent on water runoff from glaciers that melt during the warmer summer months, a continuation of the current retreat will eventually deplete the glacial ice and substantially reduce or eliminate runoff. A reduction in runoff will affect the ability to irrigate crops and will reduce summer stream flows necessary to keep dams and reservoirs replenished.

Approximately 2.4 billion people live in the drainage basin of the Himalayan rivers. India, China, Pakistan, Afghanistan, Bangladesh, Nepal and Myanmar could experience floods followed by severe droughts in coming decades. The west coast of North America, which gets much of its water from glaciers in mountain ranges such as the Rocky Mountains and Sierra Nevada, also would be affected.

Ozone and UV-B

Some scientists think agriculture could be affected by any decrease in stratospheric ozone, which could increase biologically dangerous ultraviolet radiation B. Excess ultraviolet radiation B can directly affect plant physiology and cause massive amounts of mutations, and indirectly through changed pollinator behavior, though such changes are not simple to quantify. However, it has not yet been ascertained whether an increase in greenhouse gases would decrease stratospheric ozone levels.

In addition, a possible effect of rising temperatures is significantly higher levels of ground-level ozone, which would substantially lower yields.

ENSO Effects on Agriculture

ENSO (El Niño Southern Oscillation) will affect monsoon patterns more intensely in the future as climate change warms up the ocean's water. Crops that lie on the equatorial belt or under the tropical Walker circulation, such as rice, will be affected by varying monsoon patterns and more unpredictable weather. Scheduled planting and harvesting based on weather patterns will become less effective.

Areas such as Indonesia where the main crop consists of rice will be more vulnerable to the increased intensity of ENSO effects in the future of climate change. University of Washington professor, David Battisti, researched the effects of future ENSO patterns on the Indonesian rice agriculture using [IPCC]'s 2007 annual report and 20 different logistical models mapping out climate factors such as wind pressure, sea-level, and humidity, and found that rice harvest will experience a decrease in yield. Bali and Java, which holds 55% of the rice yields in Indonesia, will be likely to experience 9–10% probably of delayed monsoon patterns, which prolongs the hungry season. Normal planting of rice crops begin in October and harvest by January. However, as climate change affects ENSO and consequently delays planting, harvesting will be late and in drier conditions, resulting in less potential yields.

Mitigation and Adaptation

In Developed Countries

Several mitigation measures for use in developed countries have been proposed:

- Breeding more resilient crop varieties, and diversification of crop species.
- Using improved agroforestry species.
- Capture and retention of rainfall, and use of improved irrigation practices.
- Increasing forest cover and Agroforestry.
- Use of emerging water harvesting techniques such as contour trenching.

In Developing Countries

The Intergovernmental Panel on Climate Change (IPCC) has reported that agriculture is responsible for over a quarter of total global greenhouse gas emissions. Given that agriculture's share in global gross domestic product (GDP) is about 4%, these figures suggest that agriculture is highly greenhouse gas intensive. Innovative agricultural practices and technologies can play a role in climate change mitigation and adaptation. This adaptation and mitigation potential is nowhere more pronounced than in developing countries where agricultural productivity remains low; poverty, vulnerability and food insecurity remain high; and the direct effects of climate change are expected to be especially harsh. Creating the necessary agricultural technologies and harnessing them to enable developing countries to adapt their agricultural systems to changing climate will require innovations in policy and institutions as well. In this context, institutions and policies are important at multiple scales.

Travis Lybbert and Daniel Sumner suggest six policy principles:

- The best policy and institutional responses will enhance information flows, incentives and flexibility.
- Policies and institutions that promote economic development and reduce poverty will often improve agricultural adaptation and may also pave the way for more effective climate change mitigation through agriculture.
- Business as usual among the world's poor is not adequate.
- Existing technology options must be made more available and accessible without overlooking complementary capacity and investments.
- Adaptation and mitigation in agriculture will require local responses, but effective policy responses must also reflect global impacts and inter-linkages.
- Trade will play a critical role in both mitigation and adaptation, but will itself be shaped importantly by climate change.

The Agricultural Model Intercomparison and Improvement Project (AgMIP) was developed in 2010 to evaluate agricultural models and intercompare their ability to predict climate impacts. In sub-Saharan Africa and South Asia, South America and East Asia, AgMIP regional research teams (RRTs) are conducting integrated assessments to improve understanding of agricultural impacts of climate change (including biophysical and economic impacts) at national and regional scales. Other AgMIP initiatives include global gridded modeling, data and information technology (IT) tool development, simulation of crop pests and diseases, site-based crop-climate sensitivity studies, and aggregation and scaling.

Impact of Agriculture on Climate Change

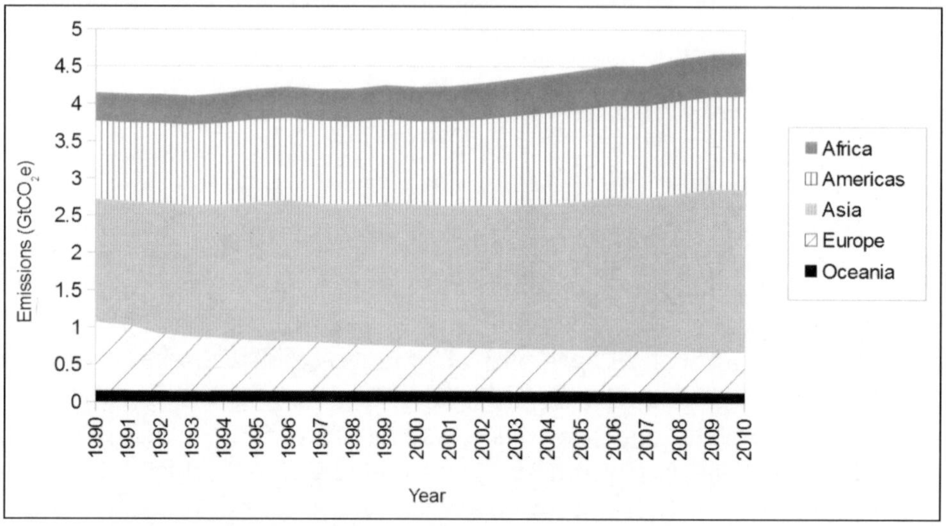

Greenhouse gas emissions from agriculture, by region, 1990-2010.

The agricultural sector is a driving force in the gas emissions and land use effects thought to cause climate change. In addition to being a significant user of land and consumer of fossil fuel, agriculture contributes directly to greenhouse gas emissions through practices such as rice production and the raising of livestock; according to the Intergovernmental Panel on Climate Change, the three main causes of the increase in greenhouse gases observed over the past 250 years have been fossil fuels, land use, and agriculture.

Land Use

Agriculture contributes to greenhouse gas increases through land use in four main ways:

- CO_2 releases linked to deforestation.
- Methane releases from rice cultivation.
- Methane releases from enteric fermentation in cattle.
- Nitrous oxide releases from fertilizer application.

Together, these agricultural processes comprise 54% of methane emissions, roughly 80% of nitrous oxide emissions, and virtually all carbon dioxide emissions tied to land use.

The planet's major changes to land cover since 1750 have resulted from deforestation in temperate regions: when forests and woodlands are cleared to make room for fields and pastures, the albedo of the affected area increases, which can result in either warming or cooling effects, depending on local conditions. Deforestation also affects regional carbon reuptake, which can result in increased concentrations of CO_2, the dominant greenhouse gas. Land-clearing methods such as slash and burn compound these effects by burning biomatter, which directly releases greenhouse gases and particulate matter such as soot into the air.

Livestock

Livestock and livestock-related activities such as deforestation and increasingly fuel-intensive farming practices are responsible for over 18% of human-made greenhouse gas emissions, including:

- 9% of global carbon dioxide emissions.
- 35–40% of global methane emissions (chiefly due to enteric fermentation and manure).
- 64% of global nitrous oxide emissions (chiefly due to fertilizer use).

Livestock activities also contribute disproportionately to land-use effects, since crops such as corn and alfalfa are cultivated in order to feed the animals.

In 2010, enteric fermentation accounted for 43% of the total greenhouse gas emissions from all agricultural activity in the world. The meat from ruminants has a higher carbon equivalent footprint than other meats or vegetarian sources of protein based on a global meta-analysis of lifecycle assessment studies. Methane production by animals, principally ruminants, is estimated 15-20% global production of methane.

Worldwide, livestock production occupies 70% of all land used for agriculture, or 30% of the land surface of the Earth. The way livestock is grazed also decides the fertility of the land in the future, not circulating grazing can lead to unhealthy soil and the expansion of livestock farms effects the habitats of local animals and has led to the drop in population of many local species from being displaced.

CLIMATE CHANGE AND ECOSYSTEMS

Rainforest ecosystems are rich in biodiversity. This is the Gambia River in Senegal's Niokolo-Koba National Park.

Future climate change is expected to affect particular ecosystems, including tundra, mangroves, coral reefs, and caves.

Unchecked global warming could affect most terrestrial ecoregions. Increasing global temperature means that ecosystems will change; some species are being forced out of their habitats (possibly to extinction) because of changing conditions, while others are flourishing. Secondary effects of global warming, such as lessened snow cover, rising sea levels, and weather changes, may influence not only human activities but also the ecosystem.

For the IPCC Fourth Assessment Report, experts assessed the literature on the impacts of climate change on ecosystems. Rosenzweig concluded that over the last three decades, human-induced warming had likely had a discernible influence on many physical and biological systems. Schneider concluded, with very high confidence, that regional temperature trends had already affected species and ecosystems around the world. With high confidence, they concluded that climate change would result in the extinction of many species and a reduction in the diversity of ecosystems.

- Terrestrial ecosystems and biodiversity: With a warming of 3 °C, relative to 1990 levels, it is likely that global terrestrial vegetation would become a net source of carbon. With high confidence, Schneider concluded that a global mean temperature increase of around 4 °C by 2100 would lead to major extinctions around the globe.

- Marine ecosystems and biodiversity: With very high confidence, Schneider concluded that a warming of 2 °C above 1990 levels would result in mass mortality of coral reefs globally. In addition, several studies dealing with planktonic organisms and modelling have shown that temperature plays a transcendental role in marine microbial food webs, which may have a deep influence on the biological carbon pump of marine planktonic pelagic and mesopelagic ecosystems.

- Freshwater ecosystems: Above about a 4 °C increase in global mean temperature by 2100, Schneider concluded, with high confidence, that many freshwater species would become extinct.

Impacts

Studying the association between Earth climate and extinctions over the past 520 million years, scientists from the University of York write, “The global temperatures predicted for the coming centuries may trigger a new ‘mass extinction event’, where over 50 percent of animal and plant species would be wiped out”.

Many of the species at risk are Arctic and Antarctic fauna such as polar bears and emperor penguins. In the Arctic, the waters of Hudson Bay are ice-free for three weeks longer than they were thirty years ago, affecting polar bears, which prefer to hunt on sea ice. Species that rely on cold weather conditions such as gyrfalcons, and snowy owls that prey on lemmings that use the cold winter to their advantage may be hit hard. Marine invertebrates enjoy peak growth at the temperatures they have adapted to, regardless of how cold these may be, and cold-blooded animals found at greater latitudes and altitudes generally grow faster to compensate for the short growing season. Warmer-than-ideal conditions result in higher metabolism and consequent reductions in body size despite increased foraging, which in turn elevates the risk of predation. Indeed, even a slight increase in temperature during development impairs growth efficiency and survival rate in rainbow trout.

Rising temperatures are beginning to have a noticeable impact on birds, and butterflies have shifted their ranges northward by 200 km in Europe and North America. Plants lag behind, and larger animals' migration is slowed down by cities and roads. In Britain, spring butterflies are appearing an average of 6 days earlier than two decades ago.

A 2002 report in *Nature* surveyed the scientific literature to find recent changes in range or seasonal behaviour by plant and animal species. Of species showing recent change, 4 out of 5 shifted their ranges towards the poles or higher altitudes, creating "refugee species". Frogs were breeding, flowers blossoming and birds migrating an average 2.3 days earlier each decade; butterflies, birds and plants moving towards the poles by 6.1 km per decade. A 2005 study concludes human activity is the cause of the temperature rise and resultant changing species behaviour, and links these effects with the predictions of climate models to provide validation for them. Scientists have observed that Antarctic hair grass is colonizing areas of Antarctica where previously their survival range was limited.

Mechanistic studies have documented extinctions due to recent climate change: McLaughlin documented two populations of Bay checkerspot butterfly being threatened by precipitation change. Parmesan states, "Few studies have been conducted at a scale that encompasses an entire species" and McLaughlin agreed "few mechanistic studies have linked extinctions to recent climate change".

Many species of freshwater and saltwater plants and animals are dependent on glacier-fed waters to ensure a cold water habitat that they have adapted to. Some species of freshwater fish need cold water to survive and to reproduce, and this is especially true with salmon and cutthroat trout. Reduced glacier runoff can lead to insufficient stream flow to allow these species to thrive. Ocean krill, a cornerstone species, prefer cold water and are the primary food source for aquatic mammals such as the blue whale. Alterations to the ocean currents, due to increased freshwater inputs from glacier melt, and the potential alterations to thermohaline circulation of the worlds oceans, may affect existing fisheries upon which humans depend as well.

The white lemuroid possum, only found in the Daintree mountain forests of northern Queensland, may be the first mammal species to be driven extinct by global warming in Australia. In 2008, the white possum has not been seen in over three years. The possums cannot survive extended temperatures over 30 °C (86 °F), which occurred in 2005.

A 27-year study of the largest colony of Magellanic penguins in the world, published in 2014, found that extreme weather caused by climate change is responsible for killing 7% of penguin chicks per year on average, and in some years studied climate change accounted for up to 50% of all chick deaths. Since 1987, the number of breeding pairs in the colony has reduced by 24%.

Climate change is leading to a mismatch between the snow camouflage of arctic animals such as snowshoe hares with the increasingly snow-free landscape.

Forests

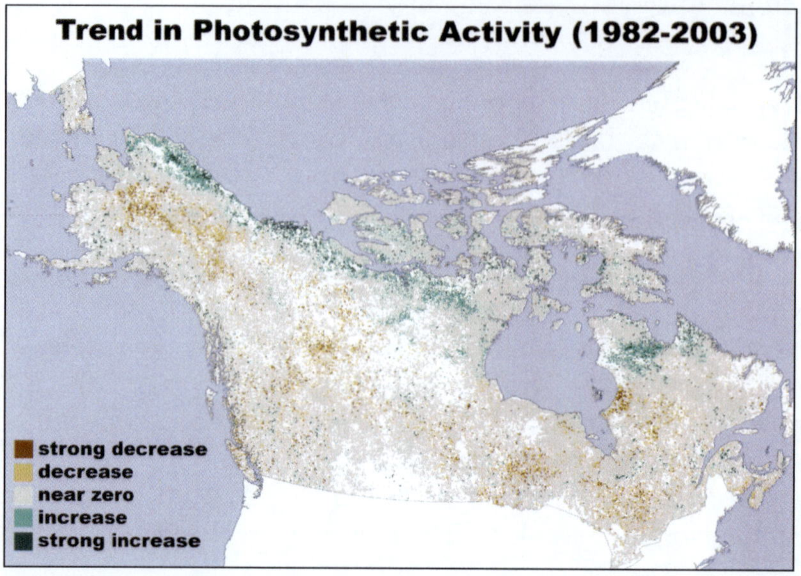

Change in Photosynthetic Activity in Northern Forests 1982-2003; NASA Earth Observatory.

Pine forests in British Columbia have been devastated by a pine beetle infestation, which has expanded unhindered since 1998 at least in part due to the lack of severe winters since that time; a few days of extreme cold kill most mountain pine beetles and have kept outbreaks in the past naturally contained. The infestation, which has killed about half of the province's lodgepole pines (33 million acres or 135,000 km²) is an order of magnitude larger than any previously recorded outbreak. One reason for unprecedented host tree mortality may be due to that the mountain pine beetles have higher reproductive success in lodgepole pine trees growing in areas where the trees have not experienced frequent beetle epidemics, which includes much of the current outbreak area. In 2007 the outbreak spread, via unusually strong winds, over the continental divide to Alberta. An epidemic also started, be it at a lower rate, in 1999 in Colorado, Wyoming, and Montana. The United States forest service predicts that between 2011 and 2013 virtually all 5 million acres (20,000 km²) of Colorado's lodgepole pine trees over five inches (127 mm) in diameter will be lost.

As the northern forests are a carbon sink, while dead forests are a major carbon source, the loss of such large areas of forest has a positive feedback on global warming. In the worst years, the carbon emission due to beetle infestation of forests in British Columbia alone approaches that of an average year of forest fires in all of Canada or five years worth of emissions from that country's transportation sources.

Wildfires

Besides the immediate ecological and economic impact, the huge dead forests provide a fire risk. Even many healthy forests appear to face an increased risk of forest fires because of warming climates. The 10-year average of boreal forest burned in North America, after several decades of around 10,000 km² (2.5 million acres), has increased steadily since 1970 to more than 28,000 km² (7 million acres) annually. Though this change may be due in part to changes in forest management practices, in the western U.S., since 1986, longer, warmer summers have resulted in a fourfold increase of major wildfires and a sixfold increase in the area of forest burned, compared to the period from 1970 to 1986. A similar increase in wildfire activity has been reported in Canada from 1920 to 1999.

Forest fires in Indonesia have dramatically increased since 1997 as well. These fires are often actively started to clear forest for agriculture. They can set fire to the large peat bogs in the region and the CO_2 released by these peat bog fires has been estimated, in an average year, to be 15% of the quantity of CO_2 produced by fossil fuel combustion.

A 2018 study found that trees grow faster due to increased carbon dioxide levels, however, the trees are also eight to twelve percent lighter and denser since 1900. The authors note, "Even though a greater volume of wood is being produced today, it now contains less material than just a few decades ago".

In 2019 unusually hot and dry weather in parts of the northern hemisphere caused massive wildfires, from the Mediterranean to – in particular – the Arctic. Climate change, by rising temperatures and shifts in precipitation patterns, is amplifying the risk of wildfires and prolonging their season. The northern part of the world is warming faster than the planet on average. The average June temperature in the parts of Siberia, where wildfires are raging, was almost ten degrees higher than the 1981–2010 average. Temperatures in Alaska reach record highs of up to 90°F (32°C) on 4 July, fuelling fires in the state, including along the Arctic Circle.

In addition to the direct threat from burning, wildfires cause air pollution, that can be carried over long distances, affecting air quality in far away regions. Wildfires also release carbon dioxide into the atmosphere, contributing to global warming. For example, the 2014 megafires in Canada burned more than 7 million acres of forest, releasing more than 103 million tonnes of carbon – half as much as all the plants in Canada typically absorb in an entire year.

Wildfires are common in the northern hemisphere between May and October, but the latitude, intensity, and the length of the fires, were particularly unusual. In June 2019, the Copernicus Atmosphere Monitoring Service (CAMS) has tracked over 100 intense and long-lived wildfires in the Arctic. In June alone, they emitted 50 megatones of carbon dioxide - equivalent to Sweden's annual GHG emissions. This is more than was released by Arctic fires in the same month in the years 2010 - 2018 combined. The fires have been most severe in Alaska and Siberia, where some cover territory equal to almost 100 000 football pitches. In Alberta, one fire was bigger than 300 000 pitches. In Alaska alone, CAMS has registered almost 400 wildfires this year, with new ones igniting every day. In Canada, smoke from massive wildfires near Ontario are producing large amounts of air pollution. The heat wave in Europe also caused wildfires in a number of countries, including Germany, Greece and Spain. The heat is drying forests and making them more susceptible to wildfires. Boreal forests are now burning at a rate unseen in at least 10,000 years.

The Arctic region, is particularly sensitive and warming faster than most other regions. Particles of smoke can land on snow and ice, causing them to absorb sunlight that it would otherwise reflect, accelerating the warming. Fires in the Arctic also increase the risk of permafrost thawing that releases methane - strong greenhouse gas. Improving forecasting systems is important to solve the problem. In view of the risks, WMO has created a Vegetation Fire and Smoke Pollution

Warning and Advisory System for forecasting fires and related impacts and hazards across the globe. WMO's Global Atmosphere Watch Programme has released a short video about the issue.

Mountains

Mountains cover approximately 25 percent of earth's surface and provide a home to more than one-tenth of global human population. Changes in global climate pose a number of potential risks to mountain habitats. Researchers expect that over time, climate change will affect mountain and lowland ecosystems, the frequency and intensity of forest fires, the diversity of wildlife, and the distribution of fresh water.

Studies suggest a warmer climate in the United States would cause lower-elevation habitats to expand into the higher alpine zone. Such a shift would encroach on the rare alpine meadows and other high-altitude habitats. High-elevation plants and animals have limited space available for new habitat as they move higher on the mountains in order to adapt to long-term changes in regional climate.

Changes in climate will also affect the depth of the mountains snowpacks and glaciers. Any changes in their seasonal melting can have powerful impacts on areas that rely on freshwater runoff from mountains. Rising temperature may cause snow to melt earlier and faster in the spring and shift the timing and distribution of runoff. These changes could affect the availability of freshwater for natural systems and human uses.

Oceans

Ocean Acidification

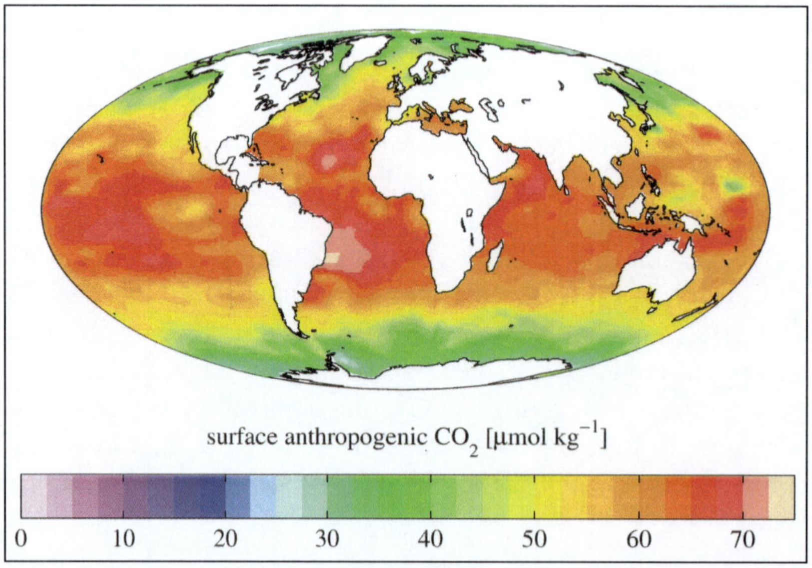

Estimated annual mean sea surface anthropogenic dissolved inorganic carbon concentration for the present day (normalised to year 2002) from the Global Ocean Data Analysis Project v2 (GLODAPv2) climatology.

Ocean acidification poses a severe threat to the earth's natural process of regulating atmospheric Co_2 levels, causing a decrease in water's ability to dissolve oxygen and created oxygen-vacant bodies of water called "dead zones." The ocean absorbs up to 55% of atmospheric carbon dioxide, lessoning the effects of climate change. This diffusion of carbon dioxide into seawater results in three acidic molecules: bicarbonate ion (HCO_3-), aqueous carbon dioxide (CO_2aq), and carbonic acid (H_2CO_3). These three compounds increase the ocean's acidity, decreasing its ph by up to 0.1 per 100ppm (part per million) of atmospheric CO_2. The increase of ocean acidity also decelerates the rate of calcification in salt water, leading to slower growing reefs which support a whopping 25% of marine life. As seen with the great barrier reef, the increase in ocean acidity in not only killing the coral, but also the wildly diverse population of marine inhabitants which coral reefs support.

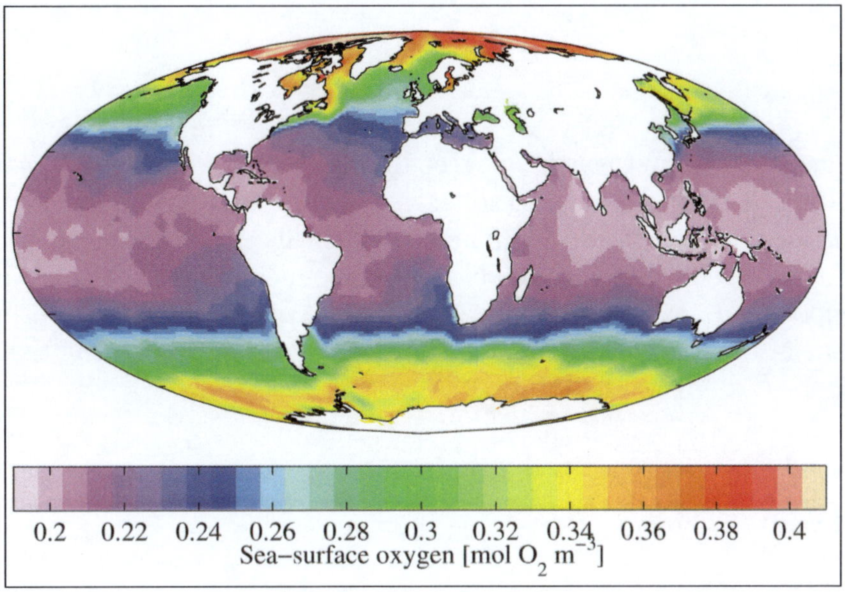

Annual mean sea surface dissolved oxygen from the World Ocean Atlas 2009. Dissolved oxygen here is in mol O_2m^{-3}.

Dissolved Oxygen

Another issue faced by increasing global temperatures is the decrease of the ocean's ability to dissolve oxygen, one with potentially more severe consequences than other repercussions of global warming. Ocean depths between 100 meters and 1,000 meters are known as "oceanic mid zones" and host a plethora of biologically diverse species, one of which being zooplankton. Zooplankton feed on smaller organisms such as phytoplankton, which are an integral part of the marine food web. Phytoplankton perform photosynthesis, receiving energy from light, and provide sustenance and energy for the larger zooplankton, which provide sustenance and energy for the even larger fish, and so on up the food chain. The increase in oceanic temperatures lowers the ocean's ability to retain oxygen generated from phytoplankton, and therefore reduces the amount of bioavailable oxygen that fish and other various marine wildlife rely on for their survival. This creates marine dead zones, and the phenomenon has already generated multiple marine dead zones around the world, as marine currents effectively "trap" the deoxygenated water.

Algal Bloom

Climate change can increase the frequency and the magnitude of algal bloom. In 2019 the biggest Sargassum bloom ever seen, create a crisis in the Tourism industry in North America. The event was probably caused by Climate Change and Fertilizers. Several Caribbean countries, even considered declare am emergency state because of the impact on tourism. The bloom can benefit the marine life, but, can also block the sunlight necessary for it.

Impact on Phytoplankton

Satellite measurement and chlorophyll observations show decline in the number of phytoplankton, microorganisms that produce half of the earth's oxygen, absorb half of the world carbon dioxide and serve foundation of the entire marine food chain. The decline is probably linked to climate change. However, there are some measurements that show increases in the number of phytoplankton.

Combined Impact

Eventually the planet will warm to such a degree that the ocean's ability to dissolve water will no longer exist, resulting in a worldwide dead zone. Dead zones, in combination with ocean acidification, will usher in an era where marine life in most forms will cease to exist, causing a sharp decline in the amount of oxygen generated through bio carbon sequestration, perpetuating the cycle. This disruption to the food chain will cascade upward, thinning out populations of primary consumers, secondary consumers, tertiary consumers, etc., as primary consumers being the initial victims of these phenomenon.

Fresh Water

Disruption to Water Cycle

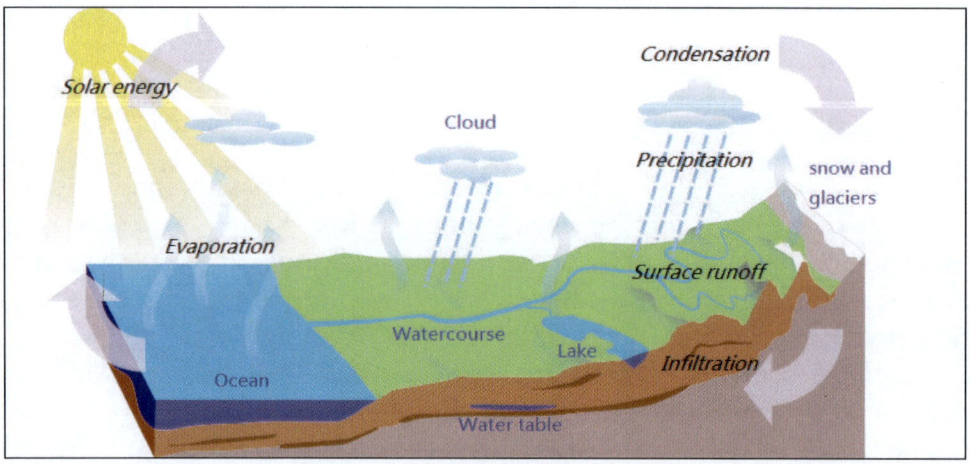

The water cycle.

Fresh water covers only 0.8% of the Earth's surface, but contains up to 6% of all life on the planet. However, the impacts climate change deal to its ecosystems are often overlooked. Very few studies showcase the potential results of climate change on large-scale ecosystems which are reliant on freshwater, such as river ecosystems, lake ecosystems, desert ecosystems, etc. However, a comprehensive study published in 2009 delves into the effects to be felt by lotic (flowing) and lentic (still) freshwater ecosystems in the American Northeast. According to the study, persistent rainfall, typically felt year round, will begin to diminish and rates of evaporation will increase, resulting in drier summers and more sporadic periods of precipitation throughout the year. Additionally, a decrease in snowfall is expected, which leads to less runoff in the spring when snow thaws and enters the watershed, resulting in lower-flowing fresh water rivers. This decrease in snowfall also leads to increased runoff during winter months, as rainfall cannot permeate the frozen ground usually covered by water-absorbing snow. These effects on the water cycle will wreak havoc for indigenous species residing in fresh water lakes and streams.

Salt Water Contamination and Cool Water Species

Species of fish living in cold or cool water can see a reduction in population of up to 50% in the majority of U.S. fresh water streams, according to most climate change models. The increase in metabolic demands due to higher water temperatures, in combination with decreasing amounts of food will be the main contributors to their decline. Additionally, many fish species (such as salmon) utilize seasonal water levels of streams as a means of reproducing, typically breeding when water flow is high and migrating to the ocean after spawning. Because snowfall is expected to be reduced due to climate change, water runoff is expected to decrease which leads to lower flowing streams, effecting the spawning of millions of salmon. To add to this, rising seas will begin to flood coastal river systems, converting them from fresh water habitats to saline environments where indigenous species will likely perish. In southeast Alaska, the sea rises by 3.96 cm/year, re-depositing sediment in various river channels and bringing salt water inland. This rise in sea level not only contaminates streams and rivers with saline water, but also the reservoirs they are connected to, where species such as Sockeye Salmon live. Although this species of Salmon can survive in both salt and fresh water, the loss of a body of fresh water stops them from reproducing in the spring, as the spawning process requires fresh water. Undoubtedly, the loss of fresh water systems of lakes and rivers in Alaska will result in the imminent demise of the state's once-abundant population of salmon.

Eagle River in central Alaska, home to various indigenous freshwater species.

Combined Impact

In general, as the planet warms, the amount of fresh water bodies across the planet decreases, as evaporation rates increase, rain patterns become more sporadic, and watershed patterns become fragmented, resulting in less cyclical water flow in river and stream systems. This disruption to fresh water cycles disrupts the feeding, mating, and migration patterns of organisms reliant on fresh water ecosystems. Additionally, the encroachment of saline water into fresh water river systems endangers indigenous species which can only survive in fresh water.

Species Migration

In 2010, a gray whale was found in the Mediterranean Sea, even though the species had not been seen in the North Atlantic Ocean since the 18th century. The whale is thought to have migrated from the Pacific Ocean via the Arctic. Climate Change & European Marine Ecosystem Research (CLAMER) has also reported that the Neodenticula seminae alga has been found in the North Atlantic, where it had gone extinct nearly 800,000 years ago. The alga has drifted from the Pacific Ocean through the Arctic, following the reduction in polar ice.

In the Siberian subarctic, species migration is contributing to another warming albedo-feedback, as needle-shedding larch trees are being replaced with dark-foliage evergreen conifers which can absorb some of the solar radiation that previously reflected off the snowpack beneath the forest canopy. It has been projected many fish species will migrate towards the North and South poles as a result of climate change, and that many species of fish near the Equator will go extinct as a result of global warming.

Migratory birds are especially at risk for endangerment due to the extreme dependability on temperature and air pressure for migration, foraging, growth, and reproduction. Much research has been done on the effects of climate change on birds, both for future predictions and for conservation. The species said to be most at risk for endangerment or extinction are populations that are not of conservation concern. It is predicted that a 3.5 degree increase in surface temperature will occur by year 2100, which could result in between 600 and 900 extinctions, which mainly will occur in the tropical environments.

Species Adaptation

In November 2019 it was revealed that a 45-year study indicated that climate change had affected the gene pool of the red deer population on Rùm, one of the Inner Hebrides islands, Scotland. Warmer temperatures resulted in deer giving birth on average three days earlier for each decade of the study. The gene which selects for earlier birth has increased in the population because does with the gene have more calves over their lifetime. Dr Timothée Bonnet, of the Australian National University, leader of the study, said they had "documented evolution in action".

In December 2019 the results of a joint study by Chicago's Field Museum and the University of Michigan into changes in the morphology of birds was published in Ecology Letters. The study uses bodies of birds which died as a result of colliding with buildings in Chicago, Illinois, since 1978. The sample is made up of over 70,000 specimins from 52 species and span the period from 1978 to 2016. The study shows that the length of birds' lower leg bones (an indicator of body sizes) shortened by an average of 2.4% and their wings lengthened by 1.3%. The findings of the study suggest the morphological changes are the result of climate change, and demonstrate an example of evolutionary change following Bergmann's rule.

Impacts of Species Degradation due to Climate Change on Livelihoods

The livelihoods of nature dependent communities depend on abundance and availability of certain species. Climate change conditions such as increase in atmospheric temperature and carbon dioxide concentration directly affect availability of biomass energy, food, fiber and other ecosystem services. Degradation of species supplying such products directly affect the livelihoods of people relying on them more so in Africa. The situation is likely to be exacerbated by changes in rainfall variability which is likely to give dominance to invasive species especially those that are spread across large latitudinal gradients. The effects that climate change has on both plant and animal species within certain ecosystems has the ability to directly affect the human inhabitants who rely on natural resources. Frequently, the extinction of plant and animal species create a cyclic relationship of species endangerment in ecosystems which are directly affected by climate change.

Agriculture

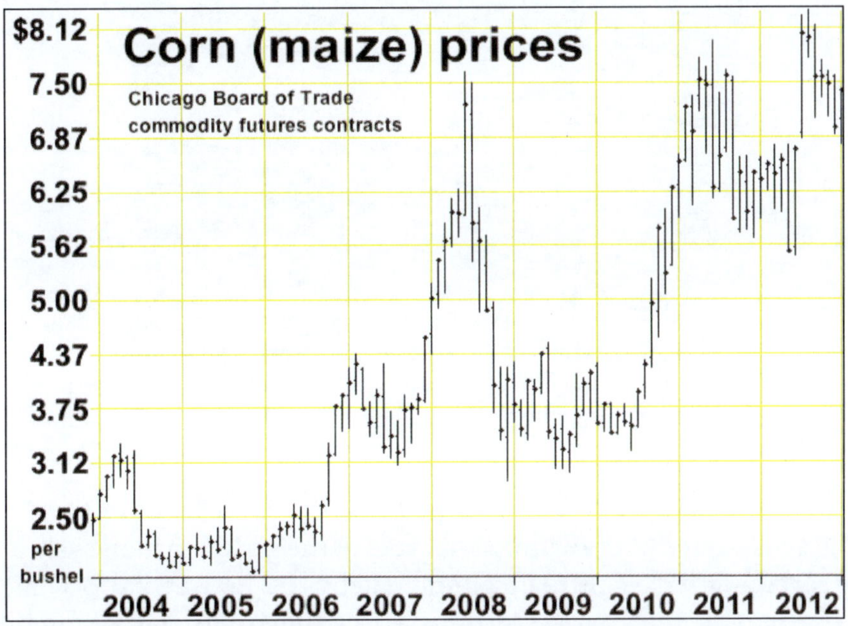

Price of corn in North America, in U.S. dollars per bushel, 2004-2012.

Droughts have been occurring more frequently because of global warming and they are expected to become more frequent and intense in Africa, southern Europe, the Middle East, most of the Americas, Australia, and Southeast Asia. Their impacts are aggravated because of increased water demand, population growth, urban expansion, and environmental protection efforts in many areas. Droughts result in crop failures and the loss of pasture grazing land for livestock.

Droughts are becoming more frequent and intense in arid and semiarid western North America as temperatures have been rising, advancing the timing and magnitude of spring snow melt floods and reducing river flow volume in summer. Direct effects of climate change include increased heat and water stress, altered crop phenology, and disrupted symbiotic interactions. These effects may be exacerbated by climate changes in river flow, and the combined effects are likely to reduce the abundance of native trees in favor of non-native herbaceous and drought-tolerant competitors, reduce the habitat quality for many native animals, and slow litter decomposition and nutrient cycling. Climate change effects on human water demand and irrigation may intensify these effects. By 2012, North American corn prices had risen to a record $8.34 per bushel in August, leaving 20 of the 211 U.S. ethanol fuel plants idle.

FISHERIES AND CLIMATE CHANGE

Rising ocean temperatures and ocean acidification are radically altering aquatic ecosystems. Climate change is modifying fish distribution and the productivity of marine and freshwater species. This has impacts on the sustainability of fisheries and aquaculture, on the livelihoods of the communities that depend on fisheries, and on the ability of the oceans to capture and store carbon (biological pump). The effect of sea level rise means that coastal fishing communities are in the front line of climate change, while changing rainfall patterns and water use impact on inland (freshwater) fisheries and aquaculture. The full relationship between fisheries and climate change is difficult to explore due to the context of each fishery and the many pathways that climate change affects.

Fishing with a lift net in Bangladesh. Coastal fishing communities in Bangladesh are vulnerable to flooding from sea-level rises.

Role of Oceans

Oceans and coastal ecosystems play an important role in the global carbon cycle and have removed about 25% of the carbon dioxide emitted by human activities between 2000 and 2007 and about half the anthropogenic CO_2 released since the start of the Industrial Revolution. Rising ocean temperatures and ocean acidification means that the capacity of the ocean carbon sink will gradually get weaker, giving rise to global concerns expressed in the Monaco and Manado Declarations. Healthy ocean ecosystems are essential for the mitigation of climate change. Coral reefs provide habitat for millions of fish species and with no change it can provoke these reefs to die.

Island with fringing reef in the Maldives. Coral reefs are dying around the world.

Impact on Fish Production

The rising ocean acidity makes it more difficult for marine organisms such as shrimps, oysters, or corals to form their shells – a process known as calcification. Many important animals, such as zooplankton, that forms the base of the marine food chain have calcium shells. Thus the entire marine food web is being altered – there are 'cracks in the food chain'. As a result, the distribution, productivity, and species composition of global fish production is changing, generating complex and inter-related impacts on oceans, estuaries, coral reefs, mangroves and sea grass beds that provide habitats and nursery areas for fish. Changing rainfall patterns and water scarcity is impacting on river and lake fisheries and aquaculture production. After the ice age about 200,000 years ago, the global air temperature has risen 3 degrees, leading to an increase in sea temperatures.

Fish catch of the global ocean is expected to decline by 6 percent by 2100 and by 11 percent in tropical zones. Diverse models predict that by 2050, the total global fish catch potential may vary by less than 10 percent depending on the trajectory of greenhouse gas emissions, but with very significant geographical variability. Decreases in both marine and terrestrial production in almost 85 percent of coastal countries analysed are predicted, varying widely in their national capacity to adapt.

Fish populations of skipjack tuna and bigeye tuna are expected to be displaced further to the east due to the effects of climate change on ocean temperatures and currents. This will shift the fishing grounds toward the Pacific islands and away from its primary owner of Melanesia, disrupting western Pacific canneries, shifting tuna production elsewhere, and having an uncertain effect on food security.

Species that are over-fished, such as the variants of Atlantic cod, are more susceptible to the effects of climate change. Over-fished populations have less size, genetic diversity, and age than other populations of fish. This makes them more susceptible to environment related stress, including those resulting from climate change. In the case of Atlantic cod located in the Baltic Sea, which are stressed close to their upper limits, this could lead to consequences related to the population's average size and growth.

Due to climate change, the distribution of zooplankton has changed. Cool water cope-pod assemblages have moved north because the waters get warmer, they have been replaced by warm water cope-pods assemblages however it has a lower biomass and certain small species. Atlantic cod require a diet of large cope-pods but because they have moved pole-wards morality rates are high and as a result the recruitment of this cod has plummeted.

Impact on Fishing Communities

Coastal and fishing populations and countries dependent on fisheries are particularly vulnerable to climate change. Low-lying countries such as the Maldives and Tuvalu are particularly vulnerable and entire communities may become the first climate refugees. Fishing communities in Bangladesh are subject not only to sea-level rise, but also flooding and increased typhoons. Fishing communities along the Mekong river produce over 1 million tons of basa fish annually and livelihoods and fish production will suffer from saltwater intrusion resulting from rising sea level and dams.

Fisherman landing his catch, Seychelles.

While climate change increases the effects of human activities, the inverse is also applicable. Human activities also increase the impact of climate change. Human activity has been linked to lake nutrition levels, which high levels are correlated to increasing vulnerability to climate change. Lake Annecy, Lake Geneva, and Lake Bourget were subject to experiments related to their zooplankton. Lake Geneva and Lake Bourget had relatively high levels of nutrients and responded at a significant level towards factors related to climate change, such as weather variability. Lake Annecy had the lowest amount of nutrition levels and responded comparatively poorly.

Fisheries and aquaculture contribute significantly to food security and livelihoods. Fish provides essential nutrition for 3 billion people and at least 50% of animal protein and minerals to 400 million people from the poorest countries. This food security is threatened by climate change and the increasing world population. Climate change changes several parameters of the fishing population: availability, stability, access, and utilization. The specific effects of climate change on these parameters will vary widely depending on the characteristics of the area, with some areas benefiting from the shift in trends and some areas being harmed based on the factors of exposure, sensitivity, and ability to respond to said changes.The lack of oxygen in warmer waters will possibly lead to the extinction of aquatic animals Worldwide food security may not change significantly, however rural and poor populations would be disproportionately and negatively affected based on this criteria, as they lack the resources and manpower to rapidly change their infrastructure and adapt. Over 500 million people in developing countries depend, directly or indirectly, on fisheries and aquaculture for their livelihoods - aquaculture is the world's fastest growing food production system, growing at 7% annually and fish products are among the most widely traded foods, with more than 37% (by volume) of world production traded internationally.

Adaptation and Mitigation

The impacts of climate change can be addressed through adaptation and mitigation. The costs and benefits of adaptation are essentially local or national, while the costs of mitigation are essentially national whereas the benefits are global. Some activities generate both mitigation and adaptation benefits, for example, the restoration of mangrove forests can protect shorelines from erosion and provide breeding grounds for fish while also sequestering carbon.

Adaptation

Several international agencies, including the World Bank and the Food and Agriculture Organization have programs to help countries and communities adapt to global warming, for example by developing policies to improve the resilience of natural resources, through assessments of risk and vulnerability, by increasing awareness of climate change impacts and strengthening key institutions, such as for weather forecasting and early warning systems. The World Development Report 2010 shows that reducing overcapacity in fishing fleets and rebuilding fish stocks can both improve resilience to climate change and increase economic returns from marine capture fisheries by US$50 billion per year, while also reducing GHG emissions by fishing fleets. Consequently, removal of subsidies on fuel for fishing can have a double benefit by reducing emissions and overfishing.

Investment in sustainable aquaculture can buffer water use in agriculture while producing food and diversifying economic activities. Algal biofuels also show potential as algae can produce 15-300 times more oil per acre than conventional crops, such as rapeseed, soybeans, or jatropha and marine algae do not require scarce freshwater. Programs such as the GEF-funded Coral Reef Targeted Research provide advice on building resilience and conserving coral reef ecosystems, while six Pacific countries recently gave a formal undertaking to protect the reefs in a biodiversity hotspot – the Coral Triangle.

Mitigation

The oceans have removed 50% of the anthropogenic CO_2, so the oceans have absorbed much of the impact of climate change. The famous White Cliffs of Dover illustrate how the ocean captures and buries carbon. These limestone cliffs are formed from the skeletons of marine plankton called coccoliths. Similarly, petroleum formation is attributed largely to marine and aquatic plankton further illustrating the key role of the oceans in carbon sequestration.

The White Cliffs of Dover.

Exactly how the oceans capture and bury CO_2 is the subject of intense research by scientists worldwide, such as the Carboocean Project. The current level of GHG emissions means that ocean acidity will continue to increase and aquatic ecosystems will continue to degrade and change. There are feedback mechanisms involved here. For example, warmer waters can absorb less CO_2, so as ocean temperatures rise some dissolved CO_2 will be released back into the atmosphere. Warming also reduces nutrient levels in the mesopelagic zone (about 200 to 1000 m deep). This in turn limits the growth of diatoms in favour of smaller phytoplankton that are poorer biological pumps of carbon. This inhibits the ability of the ocean ecosystems to sequester carbon as the oceans warm. What is clear, is that healthy ocean and coastal ecosystems are necessary to continue the vital role of the ocean carbon sinks, as indicated, for example, by the Blue Carbon assessment prepared by UNEP and the coastal carbon sinks report of IUCN and growing evidence of the role of fish biomass in the transport of carbon from surface waters to the deep ocean.

While the various carbon finance instruments include restoration of forests (REDD) and producing clean energy (emissions trading), few address the need to finance healthy ocean and aquatic ecosystems although these are essential for continued uptake of CO_2 and GHGs. The scientific basis for ocean fertilization – to produce more phytoplankton to increase the uptake of CO_2 – has been challenged, and proposals for burial of CO_2 in the deep ocean have come under criticism from environmentalists.

Over-fishing

Although there is a decline of fisheries due to climate change, a related cause for this decrease is due to over-fishing. Over-fishing exacerbates the effects of climate change by creating conditions that make a fishing population more sensitive to environmental changes. Studies show that the state of the ocean is causing fisheries to collapse, and in areas where fisheries have not yet collapsed, the amount of over-fishing that is done is having a significant impact on the industry. Over-fishing is due to having access to the open sea, it makes it very easy for people to over fish, even if it is just for fun. There is also a high demand for sea food by fishermen, as well modern technology that has increased the amount of fish caught during each trip.

If there was a specific amount of fish that people were allowed to catch then this could very well solve the problem of over fishing. This type of limit system is in place in a few countries including New Zealand, Norway, Canada, and the United States. In these countries the limit system has successfully helped in fishing industries. These types of limit systems are called Individual fishing quota. This means that the areas where this quota exist, the government has legal entity over it and in these boundaries they are entitled to utilize their ocean resources as they wish.

References

- Harunur Rashid; Leonid Polyak; Ellen Mosley-Thompson (2011). Abrupt climate change: mechanisms, patterns, and impacts. American Geophysical Union. ISBN 9780875904849. Retrieved 17 September 2013
- Predictions-future-global-climate, longcontent: scied.ucar.edu, Retrieved 3 May, 2019
- Rhines, P. B. (2006). "Sub-Arctic oceans and global climate". Weather. 61 (4): 109–118. Bibcode:2006Wthr...61..109R. Doi:10.1256/wea.223.05
- Mcguire, Bill (2002). Nicolette Linton (ed.). Raging planet: earthquakes, volcanoes, and the tectonic threat to life on earth. Hauppauge, New York: Quarto Inc. ISBN 978-0-7641-1969-9
- Nepal, S. & Shrestha, A. B. (2015). "Impact of climate change on the hydrological regime of the Indus, Ganges and Brahmaputra river basins: a review of the literature". International Journal of Water Resources Development. 31 (2): 201–218. Doi:10.1080/07900627.2015.1030494
- Milius, Susan (13 December 2017). "Worries grow that climate change will quietly steal nutrients from major food crops". Science News. Retrieved 21 January 2018

4

Global Warming

The overall increase in the normal temperature of Earth's atmosphere due to greenhouse effect is termed as global warming. Gases such as carbon dioxide, carbon monoxide, CFC, methane, etc. are responsible for causing greenhouse effect. The topics elaborated in this chapter will help in gaining a better perspective about global warming.

Global warming refers to the phenomenon of increasing average air temperatures near the surface of Earth over the past one to two centuries. Climate scientists have since the mid-20th century gathered detailed observations of various weather phenomena (such as temperatures, precipitation, and storms) and of related influences on climate (such as ocean currents and the atmosphere's chemical composition). These data indicate that Earth's climate has changed over almost every conceivable timescale since the beginning of geologic time and that the influence of human activities since at least the beginning of the Industrial Revolution has been deeply woven into the very fabric of climate change.

Giving voice to a growing conviction of most of the scientific community, the Intergovernmental Panel on Climate Change (IPCC) was formed in 1988 by the World Meteorological Organization (WMO) and the United Nations Environment Program (UNEP). In 2013 the IPCC reported that the interval between 1880 and 2012 saw an increase in global average surface temperature of approximately 0.9 °C (1.5 °F). The increase is closer to 1.1 °C (2.0 °F) when measured relative to the preindustrial (i.e., 1750–1800) mean temperature.

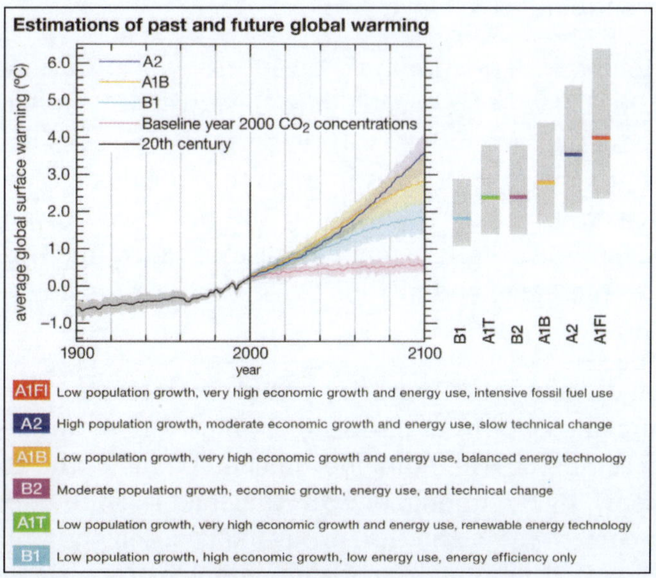

Global warming scenarios.

Graph of the predicted increase in Earth's average surface temperature according to a series of climate change scenarios that assume different levels of economic development, population growth, and fossil fuel use. The assumptions made by each scenario are given at the bottom of the graph.

A special report produced by the IPCC in 2018 honed this estimate further, noting that human beings and human activities have been responsible for a worldwide average temperature increase of between 0.8 and 1.2 °C (1.4 and 2.2 °F) of global warming since preindustrial times, and most of the warming observed over the second half of the 20th century could be attributed to human activities. It predicted that the global mean surface temperature would increase between 3 and 4 °C (5.4 and 7.2 °F) by 2100 relative to the 1986–2005 average should carbon emissions continue at their current rate. The predicted rise in temperature was based on a range of possible scenarios that accounted for future greenhouse gas emissions and mitigation (severity reduction) measures and on uncertainties in the model projections. Some of the main uncertainties include the precise role of feedback processes and the impacts of industrial pollutants known as aerosols, which may offset some warming.

Many climate scientists agree that significant societal, economic, and ecological damage would result if global average temperatures rose by more than 2 °C (3.6 °F) in such a short time. Such damage would include increased extinction of many plant and animal species, shifts in patterns of agriculture, and rising sea levels. By 2015 all but a few national governments had begun the process of instituting carbon reduction plans as part of the Paris Agreement, a treaty designed to help countries keep global warming to 1.5 °C (2.7 °F) above preindustrial levels in order to avoid the worst of the predicted effects. Authors of a special report published by the IPCC in 2018 noted that should carbon emissions continue at their present rate, the increase in average near-surface air temperatures would reach 1.5 °C sometime between 2030 and 2052. Past IPCC assessments reported that the global average sea level rose by some 19–21 cm (7.5–8.3 inches) between 1901 and 2010 and that sea levels rose faster in the second half of the 20th century than in the first half. It also predicted, again depending on a wide range of scenarios, that the global average sea level would rise 26–77 cm (10.2–30.3 inches) relative to the 1986–2005 average by 2100 for global warming of 1.5 °C, an average of 10 cm (3.9 inches) less than what would be expected if warming rose to 2 °C (3.6 °F) above preindustrial levels.

The scenarios referred to above depend mainly on future concentrations of certain trace gases, called greenhouse gases, that have been injected into the lower atmosphere in increasing amounts through the burning of fossil fuels for industry, transportation, and residential uses. Modern global warming is the result of an increase in magnitude of the so-called greenhouse effect, a warming of Earth's surface and lower atmosphere caused by the presence of water vapour, carbon dioxide, methane, nitrous oxides, and other greenhouse gases. In 2014 the IPCC reported that concentrations of carbon dioxide, methane, and nitrous oxides in the atmosphere surpassed those found in ice cores dating back 800,000 years.

Of all these gases, carbon dioxide is the most important, both for its role in the greenhouse effect and for its role in the human economy. It has been estimated that, at the beginning of the industrial age in the mid-18th century, carbon dioxide concentrations in the atmosphere were roughly 280 parts per million (ppm). By the middle of 2018 they had risen to 406 ppm, and, if fossil fuels continue to be burned at current rates, they are projected to reach 550 ppm by the mid-21st century—essentially, a doubling of carbon dioxide concentrations in 300 years.

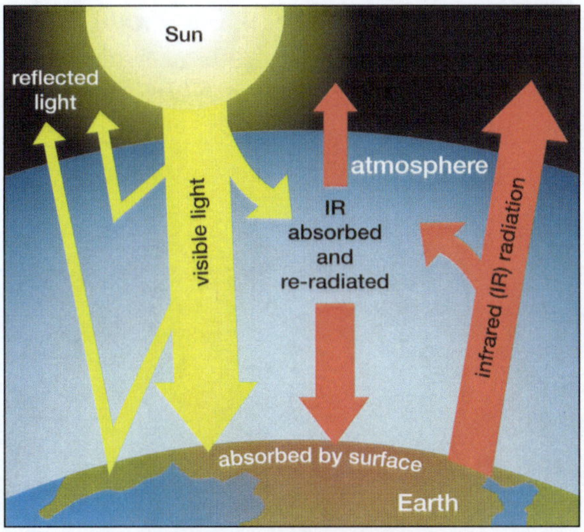

Greenhouse effect on Earth.

The greenhouse effect on Earth. Some incoming sunlight is reflected by Earth's atmosphere and surface, but most is absorbed by the surface, which is warmed. Infrared (IR) radiation is then emitted from the surface. Some IR radiation escapes to space, but some is absorbed by the atmosphere's greenhouse gases (especially water vapour, carbon dioxide, and methane) and reradiated in all directions, some to space and some back toward the surface, where it further warms the surface and the lower atmosphere.

A vigorous debate is in progress over the extent and seriousness of rising surface temperatures, the effects of past and future warming on human life, and the need for action to reduce future warming and deal with its consequences. This topic provides an overview of the scientific background and public policy debate related to the subject of global warming. It considers the causes of rising near-surface air temperatures, the influencing factors, the process of climate research and forecasting, the possible ecological and social impacts of rising temperatures, and the public policy developments since the mid-20th century.

Climatic Variation since the last Glaciation

Global warming is related to the more general phenomenon of climate change, which refers to changes in the totality of attributes that define climate. In addition to changes in air temperature, climate change involves changes to precipitation patterns, winds, ocean currents, and other measures of Earth's climate. Normally, climate change can be viewed as the combination of various natural forces occurring over diverse timescales. Since the advent of human civilization, climate change has involved an "anthropogenic," or exclusively human-caused, element, and this anthropogenic element has become more important in the industrial period of the past two centuries. The term *global warming* is used specifically to refer to any warming of near-surface air during the past two centuries that can be traced to anthropogenic causes.

To define the concepts of global warming and climate change properly, it is first necessary to recognize that the climate of Earth has varied across many timescales, ranging from an individual human life span to billions of years. This variable climate history is typically classified in terms of "regimes" or "epochs." For instance, the Pleistocene glacial epoch (about 2,600,000 to 11,700 years ago) was marked by substantial variations in the global extent of glaciers and ice sheets. These variations took place on timescales of tens to hundreds of millennia and were driven by changes in the distribution of solar radiation across Earth's surface. The distribution of solar radiation is known as the insolation pattern, and it is strongly affected by the geometry of Earth's orbit around the Sun and by the orientation, or tilt, of Earth's axis relative to the direct rays of the Sun.

A series of photographs of the Grinnell Glacier taken from the summit of Mount Gould in Glacier National Park, Montana, in 1938, 1981, 1998, and 2006 (from left to right). In 1938 the Grinnell Glacier filled the entire area at the bottom of the image. By 2006 it had largely disappeared from this view.

Worldwide, the most recent glacial period, or ice age, culminated about 21,000 years ago in what is often called the Last Glacial Maximum. During this time, continental ice sheets extended well into the middle latitude regions of Europe and North America, reaching as far south as present-day London and New York City. Global annual mean temperature appears to have been about 4–5 °C (7–9 °F) colder than in the mid-20th century. It is important to remember that these figures are a global average. In fact, during the height of this last ice age, Earth's climate was characterized by greater cooling at higher latitudes (that is, toward the poles) and relatively little cooling over large parts of the tropical oceans (near the Equator). This glacial interval terminated abruptly about 11,700 years ago and was followed by the subsequent relatively ice-free period known as the Holocene Epoch. The modern period of Earth's history is conventionally defined as residing within the Holocene. However, some scientists have argued that the Holocene Epoch terminated in the relatively recent past and that Earth currently resides in a climatic interval that could justly be called the Anthropocene Epoch—that is, a period during which humans have exerted a dominant influence over climate.

Though less dramatic than the climate changes that occurred during the Pleistocene Epoch, significant variations in global climate have nonetheless taken place over the course of the Holocene. During the early Holocene, roughly 9,000 years ago, atmospheric circulation and precipitation patterns appear to have been substantially different from those of today. For example, there is evidence for relatively wet conditions in what is now the Sahara Desert. The change from one climatic regime to another was caused by only modest changes in the pattern of insolation within the Holocene interval as well as the interaction of these patterns with large-scale climate phenomena such as monsoons and El Niño/Southern Oscillation (ENSO).

During the middle Holocene, some 5,000–7,000 years ago, conditions appear to have been relatively warm—indeed, perhaps warmer than today in some parts of the world and during certain seasons. For this reason, this interval is sometimes referred to as the Mid-Holocene Climatic

Optimum. The relative warmth of average near-surface air temperatures at this time, however, is somewhat unclear. Changes in the pattern of insolation favoured warmer summers at higher latitudes in the Northern Hemisphere, but these changes also produced cooler winters in the Northern Hemisphere and relatively cool conditions year-round in the tropics. Any overall hemispheric or global mean temperature changes thus reflected a balance between competing seasonal and regional changes. In fact, recent theoretical climate model studies suggest that global mean temperatures during the middle Holocene were probably 0.2–0.3 °C (0.4–0.5 °F) colder than average late 20th-century conditions.

Over subsequent millennia, conditions appear to have cooled relative to middle Holocene levels. This period has sometimes been referred to as the "Neoglacial". In the middle latitudes this cooling trend was associated with intermittent periods of advancing and retreating mountain glaciers reminiscent of (though far more modest than) the more substantial advance and retreat of the major continental ice sheets of the Pleistocene climate epoch.

Causes of Global Warming

Radiative Forcing

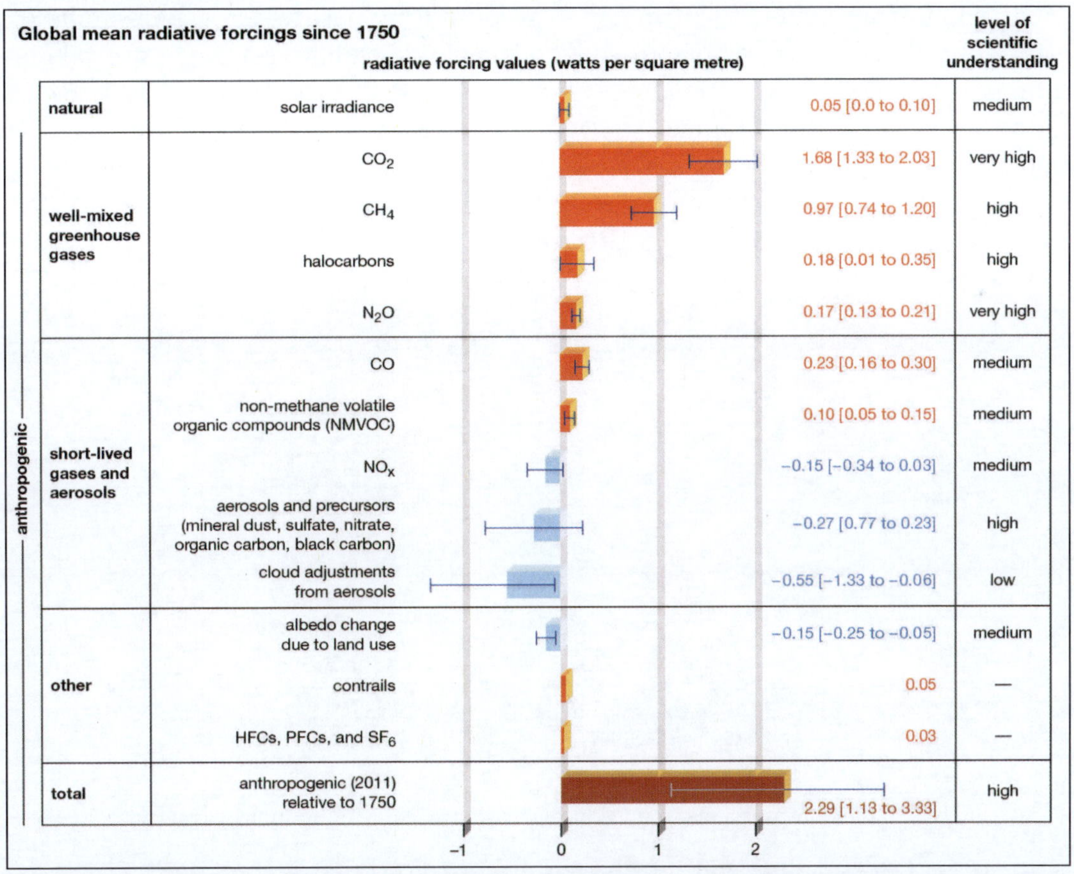

Since 1750 the concentration of carbon dioxide and other greenhouse gases has increased in Earth's atmosphere. As a result of these and other factors, Earth's atmosphere retains more heat than in the past.

It is apparent that the temperature of Earth's surface and lower atmosphere may be modified in three ways: (1) through a net increase in the solar radiation entering at the top of Earth's atmosphere, (2) through a change in the fraction of the radiation reaching the surface, and (3) through a change in the concentration of greenhouse gases in the atmosphere. In each case the changes can be thought of in terms of "radiative forcing". As defined by the IPCC, radiative forcing is a measure of the influence a given climatic factor has on the amount of downward-directed radiant energy impinging upon Earth's surface. Climatic factors are divided between those caused primarily by human activity (such as greenhouse gas emissions and aerosol emissions) and those caused by natural forces (such as solar irradiance); then, for each factor, so-called forcing values are calculated for the time period between 1750 and the present day. "Positive forcing" is exerted by climatic factors that contribute to the warming of Earth's surface, whereas "negative forcing" is exerted by factors that cool Earth's surface.

On average, about 342 watts of solar radiation strike each square metre of Earth's surface per year, and this quantity can in turn be related to a rise or fall in Earth's surface temperature. Temperatures at the surface may also rise or fall through a change in the distribution of terrestrial radiation (that is, radiation emitted by Earth) within the atmosphere. In some cases, radiative forcing has a natural origin, such as during explosive eruptions from volcanoes where vented gases and ash block some portion of solar radiation from the surface. In other cases, radiative forcing has an anthropogenic, or exclusively human, origin. For example, anthropogenic increases in carbon dioxide, methane, and nitrous oxide are estimated to account for 2.3 watts per square metre of positive radiative forcing. When all values of positive and negative radiative forcing are taken together and all interactions between climatic factors are accounted for, the total net increase in surface radiation due to human activities since the beginning of the Industrial Revolution is 1.6 watts per square metre.

Influences of Human Activity on Climate

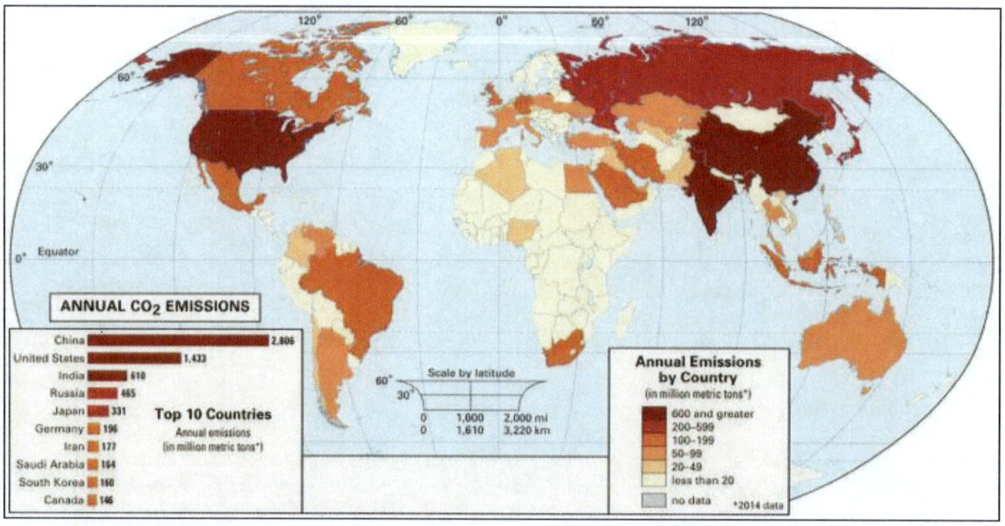

Carbon dioxide emissions Map of annual carbon dioxide emissions by country.

Human activity has influenced global surface temperatures by changing the radiative balance governing the Earth on various timescales and at varying spatial scales. The most profound and well-known anthropogenic influence is the elevation of concentrations of greenhouse gases in the atmosphere. Humans also influence climate by changing the concentrations of aerosols and ozone and by modifying the land cover of Earth's surface.

Petroleum refinery at ras tanura. Saudi Arabia.

Natural gas facility near kursk, Russia.

Water Vapour

Water vapour is the most potent of the greenhouse gases in Earth's atmosphere, but its behaviour is fundamentally different from that of the other greenhouse gases. The primary role of water vapour is not as a direct agent of radiative forcing but rather as a climate feedback—that is, as a response within the climate system that influences the system's continued activity. This distinction arises from the fact that the amount of water vapour in the atmosphere cannot, in general, be directly modified by human behaviour but is instead set by air temperatures. The warmer the surface, the greater the evaporation rate of water from the surface. As a result, increased evaporation leads to a greater concentration of water vapour in the lower atmosphere capable of absorbing longwave radiation and emitting it downward.

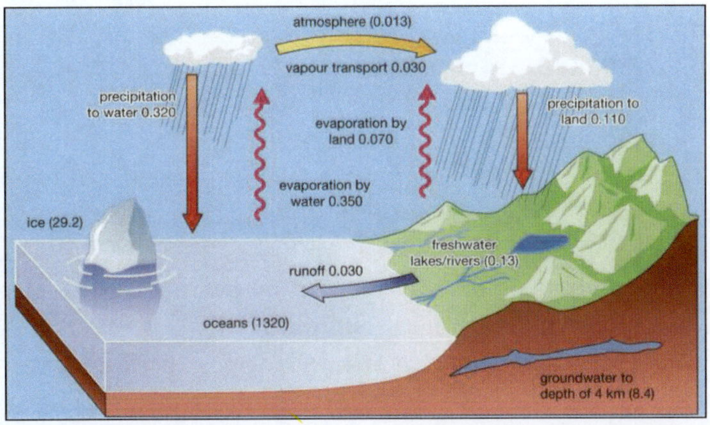

The present-day surface hydrologic cycle, in which water is transferred from the oceans through the atmosphere to the continents and back to the oceans over and beneath the land surface. The values in parentheses following the various forms of water (e.g., ice) refer to volumes in millions of cubic kilometres; those following the processes (e.g., precipitation) refer to their fluxes in millions of cubic kilometres of water per year.

Carbon Dioxide

Of the greenhouse gases, carbon dioxide (CO_2) is the most significant. Natural sources of atmospheric CO_2 include outgassing from volcanoes, the combustion and natural decay of organic matter, and respiration by aerobic (oxygen-using) organisms. These sources are balanced, on average, by a set of physical, chemical, or biological processes, called "sinks," that tend to remove CO_2 from the atmosphere. Significant natural sinks include terrestrial vegetation, which takes up CO_2 during the process of photosynthesis.

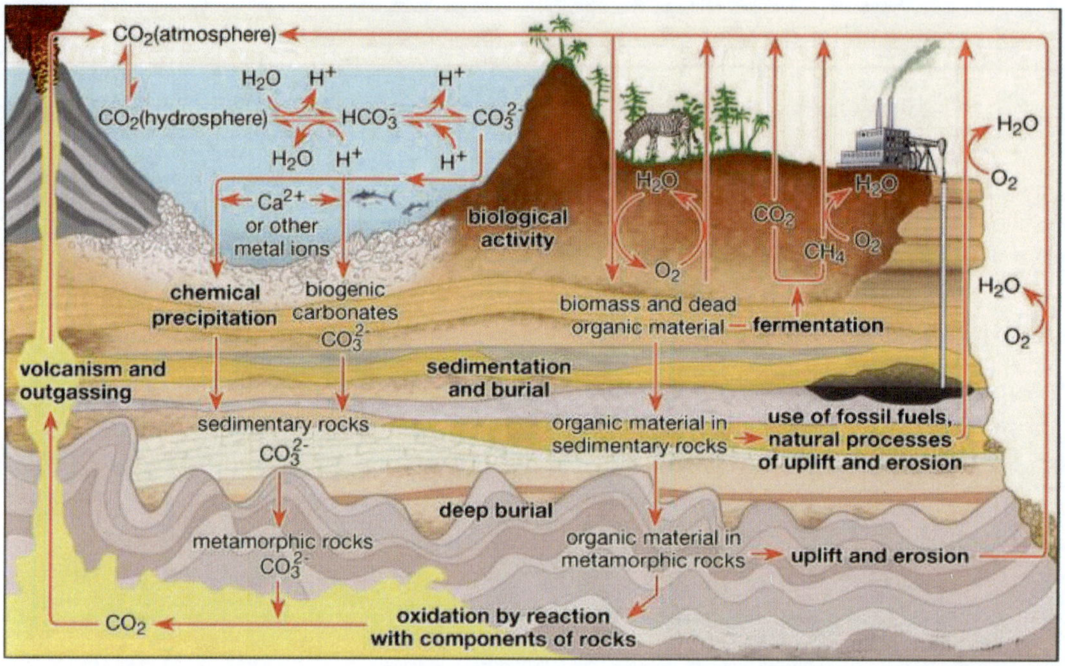

The carbon cycle Carbon is transported in various forms through the atmosphere, the hydrosphere, and geologic formations. One of the primary pathways for the exchange of carbon dioxide (CO_2) takes place between the atmosphere and the oceans; there a fraction of the CO_2 combines with water, forming carbonic acid (H_2CO_3) that subsequently loses hydrogen ions (H^+) to form bicarbonate (HCO_3^-) and carbonate (CO_3^{2-}) ions. Mollusk shells or mineral precipitates that form by the reaction of calcium or other metal ions with carbonate may become buried in geologic strata and eventually release CO_2 through volcanic outgassing. Carbon dioxide also exchanges through photosynthesis in plants and through respiration in animals. Dead and decaying organic matter may ferment and release CO_2 or methane (CH_4) or may be incorporated into sedimentary rock, where it is converted to fossil fuels. Burning of hydrocarbon fuels returns CO_2 and water (H_2O) to the atmosphere. The biological and anthropogenic pathways are much faster than the geochemical pathways and, consequently, have a greater impact on the composition and temperature of the atmosphere.

A number of oceanic processes also act as carbon sinks. One such process, called the "solubility pump," involves the descent of surface seawater containing dissolved CO_2. Another process, the "biological pump," involves the uptake of dissolved CO_2 by marine vegetation and phytoplankton (small free-floating photosynthetic organisms) living in the upper ocean or by other marine organisms that use CO_2 to build skeletons and other structures made of calcium carbonate ($CaCO_3$). As these organisms expire and fall to the ocean floor, the carbon they contain is transported downward and eventually buried at depth. A long-term balance between these natural sources and sinks leads to the background, or natural, level of CO_2 in the atmosphere.

In contrast, human activities increase atmospheric CO_2 levels primarily through the burning of fossil fuels—principally oil and coal and secondarily natural gas, for use in transportation, heating, and the generation of electrical power—and through the production of cement. Other anthropogenic sources include the burning of forests and the clearing of land. Anthropogenic emissions currently account for the annual release of about 7 gigatons (7 billion tons) of carbon into the atmosphere. Anthropogenic emissions are equal to approximately 3 percent of the total emissions of CO_2 by natural sources, and this amplified carbon load from human activities far exceeds the offsetting capacity of natural sinks (by perhaps as much as 2–3 gigatons per year).

Deforestation Smoldering remains of a plot of deforested land in the Amazon Rainforest of Brazil. Annually, it is estimated that net global deforestation accounts for about two gigatons of carbon emissions to the atmosphere.

CO_2 consequently accumulated in the atmosphere at an average rate of 1.4 ppm per year between 1959 and 2006 and roughly 2.0 ppm per year between 2006 and 2018. Overall, this rate of accumulation has been linear (that is, uniform over time). However, certain current sinks, such as the oceans, could become sources in the future. This may lead to a situation in which the concentration of atmospheric CO_2 builds at an exponential rate (that is, its rate of increase is also increasing).

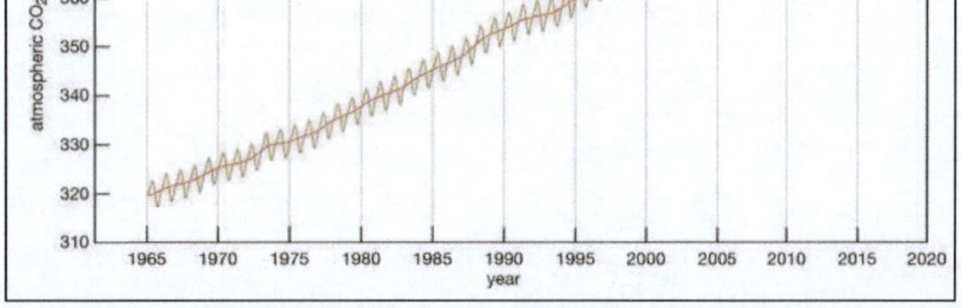

The Keeling Curve, named after American climate scientist Charles David Keeling, tracks changes in the concentration of carbon dioxide (CO_2) in Earth's atmosphere at a research station on Mauna Loa in Hawaii. Although these concentrations experience small seasonal fluctuations, the overall trend shows that CO_2 is increasing in the atmosphere.

The natural background level of carbon dioxide varies on timescales of millions of years because of slow changes in outgassing through volcanic activity. For example, roughly 100 million years ago, during the Cretaceous Period (145 million to 66 million years ago), CO_2 concentrations appear to have been several times higher than they are today (perhaps close to 2,000 ppm). Over the past 700,000 years, CO_2 concentrations have varied over a far smaller range (between roughly 180 and 300 ppm) in association with the same Earth orbital effects linked to the coming and going of the Pleistocene ice ages. By the early 21st century, CO_2 levels had reached 384 ppm, which is approximately 37 percent above the natural background level of roughly 280 ppm that existed at the beginning of the Industrial Revolution. Atmospheric CO_2 levels continued to increase, and by 2018 they had reached 410 ppm. Such levels are believed to be the highest in at least 800,000 years according to ice core measurements and may be the highest in at least 5 million years according to other lines of evidence.

Radiative forcing caused by carbon dioxide varies in an approximately logarithmic fashion with the concentration of that gas in the atmosphere. The logarithmic relationship occurs as the result of a saturation effect wherein it becomes increasingly difficult, as CO_2 concentrations increase, for additional CO_2 molecules to further influence the "infrared window" (a certain narrow band of wavelengths in the infrared region that is not absorbed by atmospheric gases). The logarithmic relationship predicts that the surface warming potential will rise by roughly the same amount for each doubling of CO_2 concentration. At current rates of fossil fuel use, a doubling of CO_2 concentrations over preindustrial levels is expected to take place by the middle of the 21st century (when CO_2 concentrations are projected to reach 560 ppm). A doubling of CO_2 concentrations would represent an increase of roughly 4 watts per square metre of radiative forcing. Given typical estimates of "climate sensitivity" in the absence of any offsetting factors, this energy increase would lead to a warming of 2 to 5 °C (3.6 to 9 °F) over preindustrial times. The total radiative forcing by anthropogenic CO_2 emissions since the beginning of the industrial age is approximately 1.66 watts per square metre.

Methane

Methane (CH_4) is the second most important greenhouse gas. CH_4 is more potent than CO_2 because the radiative forcing produced per molecule is greater. In addition, the infrared window is less saturated in the range of wavelengths of radiation absorbed by CH_4, so more molecules may fill in the region. However, CH_4 exists in far lower concentrations than CO_2 in the atmosphere, and its concentrations by volume in the atmosphere are generally measured in parts per billion (ppb) rather than ppm. CH_4 also has a considerably shorter residence time in the atmosphere than CO_2 (the residence time for CH_4 is roughly 10 years, compared with hundreds of years for CO_2).

Natural sources of methane include tropical and northern wetlands, methane-oxidizing bacteria that feed on organic material consumed by termites, volcanoes, seepage vents of the seafloor in regions rich with organic sediment, and methane hydrates trapped along the continental shelves of the oceans and in polar permafrost. The primary natural sink for methane is the atmosphere itself, as methane reacts readily with the hydroxyl radical (·OH) within the troposphere to form CO_2 and water vapour (H_2O). When CH_4 reaches the stratosphere, it is destroyed. Another natural sink is soil, where methane is oxidized by bacteria.

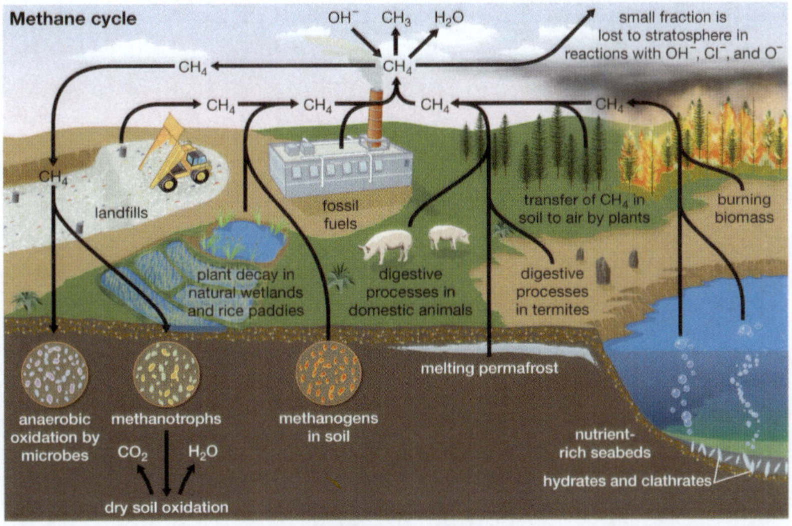

Methane cycle.

As with CO_2, human activity is increasing the CH_4 concentration faster than it can be offset by natural sinks. Anthropogenic sources currently account for approximately 70 percent of total annual emissions, leading to substantial increases in concentration over time. The major anthropogenic sources of atmospheric CH_4 are rice cultivation, livestock farming, the burning of coal and natural gas, the combustion of biomass, and the decomposition of organic matter in landfills. Future trends are particularly difficult to anticipate. This is in part due to an incomplete understanding of the climate feedbacks associated with CH_4 emissions. In addition it is difficult to predict how, as human populations grow, possible changes in livestock raising, rice cultivation, and energy utilization will influence CH_4 emissions.

It is believed that a sudden increase in the concentration of methane in the atmosphere was responsible for a warming event that raised average global temperatures by 4–8 °C (7.2–14.4 °F) over a few thousand years during the so-called Paleocene-Eocene Thermal Maximum, or PETM. This episode took place roughly 55 million years ago, and the rise in CH_4 appears to have been related to a massive volcanic eruption that interacted with methane-containing flood deposits. As a result, large amounts of gaseous CH_4 were injected into the atmosphere. It is difficult to know precisely how high these concentrations were or how long they persisted. At very high concentrations, residence times of CH_4 in the atmosphere can become much greater than the nominal 10-year residence time that applies today. Nevertheless, it is likely that these concentrations reached several ppm during the PETM.

Methane concentrations have also varied over a smaller range (between roughly 350 and 800 ppb) in association with the Pleistocene ice age cycles. Preindustrial levels of CH_4 in the atmosphere were approximately 700 ppb, whereas levels exceeded 1,867 ppb in late 2018. (These concentrations are well above the natural levels observed for at least the past 650,000 years.) The net radiative forcing by anthropogenic CH_4 emissions is approximately 0.5 watt per square metre—or roughly one-third the radiative forcing of CO_2.

Surface-level Ozone and other Compounds

The next most significant greenhouse gas is surface, or low-level, ozone (O_3). Surface O_3 is a result of air pollution; it must be distinguished from naturally occurring stratospheric O_3, which has a very different role in the planetary radiation balance. The primary natural source of surface O_3 is the subsidence of stratospheric O_3 from the upper atmosphere. In contrast, the primary anthropogenic source of surface O_3 is photochemical reactions involving the atmospheric pollutant carbon monoxide (CO). The best estimates of the natural concentration of surface O_3 are 10 ppb, and the net radiative forcing due to anthropogenic emissions of surface O_3 is approximately 0.35 watt per square metre. Ozone concentrations can rise above unhealthy levels (that is, conditions where concentrations meet or exceed 70 ppb for eight hours or longer) in cities prone to photochemical smog.

Nitrous Oxides and Fluorinated Gases

Additional trace gases produced by industrial activity that have greenhouse properties include nitrous oxide (N_2O) and fluorinated gases (halocarbons), the latter including sulfur hexafluoride, hydrofluorocarbons (HFCs), and perfluorocarbons (PFCs). Nitrous oxide is responsible for 0.16 watt per square metre radiative forcing, while fluorinated gases are collectively responsible for 0.34 watt per square metre. Nitrous oxides have small background concentrations due to natural biological reactions in soil and water, whereas the fluorinated gases owe their existence almost entirely to industrial sources.

Aerosols

The production of aerosols represents an important anthropogenic radiative forcing of climate. Collectively, aerosols block—that is, reflect and absorb—a portion of incoming solar radiation, and this creates a negative radiative forcing. Aerosols are second only to greenhouse gases in relative importance in their impact on near-surface air temperatures. Unlike the decade-long residence times of the "well-mixed" greenhouse gases, such as CO_2 and CH_4, aerosols are readily flushed out of the atmosphere within days, either by rain or snow (wet deposition) or by settling out of the air (dry deposition). They must therefore be continually generated in order to produce a steady effect on radiative forcing. Aerosols have the ability to influence climate directly by absorbing or reflecting incoming solar radiation, but they can also produce indirect effects on climate by modifying cloud formation or cloud properties. Most aerosols serve as condensation nuclei (surfaces upon which water vapour can condense to form clouds); however, darker-coloured aerosols may hinder cloud formation by absorbing sunlight and heating up the surrounding air. Aerosols can be transported thousands of kilometres from their sources of origin by winds and upper-level circulation in the atmosphere.

Perhaps the most important type of anthropogenic aerosol in radiative forcing is sulfate aerosol. It is produced from sulfur dioxide (SO_2) emissions associated with the burning of coal and oil. Since the late 1980s, global emissions of SO_2 have decreased from about 151.5 million tonnes (167.0 million tons) to less than 100 million tonnes (110.2 million tons) of sulfur per year.

Nitrate aerosol is not as important as sulfate aerosol, but it has the potential to become a significant source of negative forcing. One major source of nitrate aerosol is smog (the combination of ozone with oxides of nitrogen in the lower atmosphere) released from the incomplete burning of fuel in internal-combustion engines. Another source is ammonia (NH_3), which is often used in fertilizers or released by the burning of plants and other organic materials. If greater amounts of atmospheric nitrogen are converted to ammonia and agricultural ammonia emissions continue to increase as projected, the influence of nitrate aerosols on radiative forcing is expected to grow.

Both sulfate and nitrate aerosols act primarily by reflecting incoming solar radiation, thereby reducing the amount of sunlight reaching the surface. Most aerosols, unlike greenhouse gases, impart a cooling rather than warming influence on Earth's surface. One prominent exception is carbonaceous aerosols such as carbon black or soot, which are produced by the burning of fossil fuels and biomass. Carbon black tends to absorb rather than reflect incident solar radiation, and so it has a warming impact on the lower atmosphere, where it resides. Because of its absorptive properties, carbon black is also capable of having an additional indirect effect on climate. Through its deposition in snowfall, it can decrease the albedo of snow cover. This reduction in the amount of solar radiation reflected back to space by snow surfaces creates a minor positive radiative forcing.

Natural forms of aerosol include windblown mineral dust generated in arid and semiarid regions and sea salt produced by the action of waves breaking in the ocean. Changes to wind patterns as a result of climate modification could alter the emissions of these aerosols. The influence of climate change on regional patterns of aridity could shift both the sources and the destinations of dust clouds. In addition, since the concentration of sea salt aerosol, or sea aerosol, increases with the strength of the winds near the ocean surface, changes in wind speed due to global warming and climate change could influence the concentration of sea salt aerosol. For example, some studies suggest that climate change might lead to stronger winds over parts of the North Atlantic Ocean. Areas with stronger winds may experience an increase in the concentration of sea salt aerosol.

Other natural sources of aerosols include volcanic eruptions, which produce sulfate aerosol, and biogenic sources (e.g., phytoplankton), which produce dimethyl sulfide (DMS). Other important biogenic aerosols, such as terpenes, are produced naturally by certain kinds of trees or other plants. For example, the dense forests of the Blue Ridge Mountains of Virginia in the United States emit terpenes during the summer months, which in turn interact with the high humidity and warm temperatures to produce a natural photochemical smog. Anthropogenic pollutants such as nitrate and ozone, both of which serve as precursor molecules for the generation of biogenic aerosol, appear to have increased the rate of production of these aerosols severalfold. This process appears to be responsible for some of the increased aerosol pollution in regions undergoing rapid urbanization.

Human activity has greatly increased the amount of aerosol in the atmosphere compared with the background levels of preindustrial times. In contrast to the global effects of greenhouse gases, the impact of anthropogenic aerosols is confined primarily to the Northern Hemisphere, where most of the world's industrial activity occurs. The pattern of increases in anthropogenic aerosol over time is also somewhat different from that of greenhouse gases. During the middle of the 20th century, there was a substantial increase in aerosol emissions. This appears to have been at least partially responsible for a cessation of surface warming that took place in the Northern Hemisphere from the 1940s through the 1970s. Since that time, aerosol emissions have leveled off due to antipollution measures undertaken in the industrialized countries since the 1960s. Aerosol emissions may rise in the future, however, as a result of the rapid emergence of coal-fired electric power generation in China and India.

The total radiative forcing of all anthropogenic aerosols is approximately −1.2 watts per square metre. Of this total, −0.5 watt per square metre comes from direct effects (such as the reflection of solar energy back into space), and −0.7 watt per square metre comes from indirect effects (such as the influence of aerosols on cloud formation). This negative radiative forcing represents an offset of roughly 40 percent from the positive radiative forcing caused by human activity. However, the relative uncertainty in aerosol radiative forcing (approximately 90 percent) is much greater than that of greenhouse gases. In addition, future emissions of aerosols from human activities, and the influence of these emissions on future climate change, are not known with any certainty. Nevertheless, it can be said that, if concentrations of anthropogenic aerosols continue to decrease as they have since the 1970s, a significant offset to the effects of greenhouse gases will be reduced, opening future climate to further warming.

Land-use Change

There are a number of ways in which changes in land use can influence climate. The most direct influence is through the alteration of Earth's albedo, or surface reflectance. For example, the replacement of forest by cropland and pasture in the middle latitudes over the past several centuries has led to an increase in albedo, which in turn has led to greater reflection of incoming solar radiation in those regions. This replacement of forest by agriculture has been associated with a change in global average radiative forcing of approximately −0.2 watt per square metre since 1750. In Europe and other major agricultural regions, such land-use conversion began more than 1,000 years ago and has proceeded nearly to completion. For Europe, the negative radiative forcing due to land-use change has probably been substantial, perhaps approaching −5 watts per square metre. The influence of early land use on radiative forcing may help to explain a long period of cooling in Europe that followed a period of relatively mild conditions roughly 1,000 years ago. It is generally believed that the mild temperatures of this "medieval warm period," which was followed by a long period of cooling, rivaled those of 20th-century Europe.

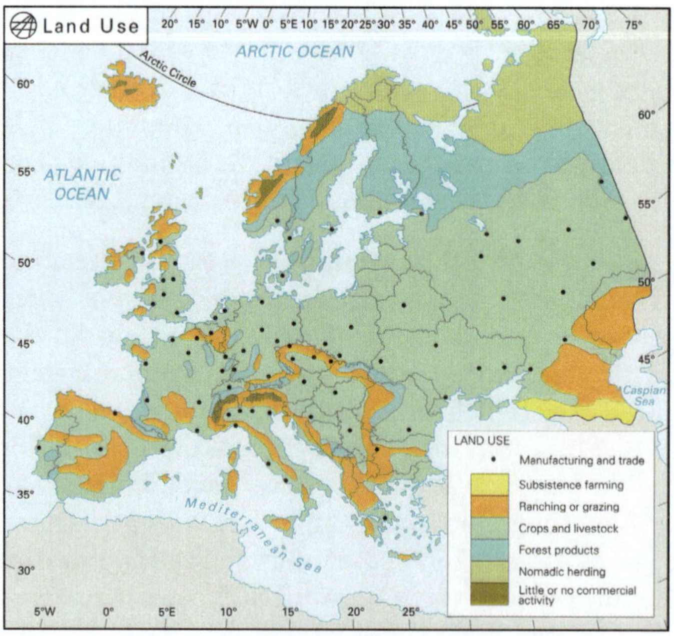

Land use in Europe.

Land-use changes can also influence climate through their influence on the exchange of heat between Earth's surface and the atmosphere. For example, vegetation helps to facilitate the evaporation of water into the atmosphere through evapotranspiration. In this process, plants take up liquid water from the soil through their root systems. Eventually this water is released through transpiration into the atmosphere, as water vapour through the stomata in leaves. While deforestation generally leads to surface cooling due to the albedo factor, the land surface may also be warmed as a result of the release of latent heat by the evapotranspiration process. The relative importance of these two factors, one exerting a cooling effect and the other a warming effect, varies by both season and region. While the albedo effect is likely to dominate in middle latitudes, especially during the period from autumn through spring, the evapotranspiration effect may dominate during the summer in the midlatitudes and year-round in the tropics. The latter case is particularly important in assessing the potential impacts of continued tropical deforestation.

The rate at which tropical regions are deforested is also relevant to the process of carbon sequestration, the long-term storage of carbon in underground cavities and biomass rather than in the atmosphere. By removing carbon from the atmosphere, carbon sequestration acts to mitigate global warming. Deforestation contributes to global warming, as fewer plants are available to take up carbon dioxide from the atmosphere. In addition, as fallen trees, shrubs, and other plants are burned or allowed to slowly decompose, they release as carbon dioxide the carbon they stored during their lifetimes. Furthermore, any land-use change that influences the amount, distribution, or type of vegetation in a region can affect the concentrations of biogenic aerosols, though the impact of such changes on climate is indirect and relatively minor.

Stratospheric Ozone Depletion

Since the 1970s the loss of ozone (O_3) from the stratosphere has led to a small amount of negative radiative forcing of the surface. This negative forcing represents a competition between two distinct effects caused by the fact that ozone absorbs solar radiation. In the first case, as ozone levels in the stratosphere are depleted, more solar radiation reaches Earth's surface. In the absence of any other influence, this rise in insolation would represent a positive radiative forcing of the surface. However, there is a second effect of ozone depletion that is related to its greenhouse properties. As the amount of ozone in the stratosphere is decreased, there is also less ozone to absorb longwave radiation emitted by Earth's surface. With less absorption of radiation by ozone, there is a corresponding decrease in the downward reemission of radiation. This second effect overwhelms the first and results in a modest negative radiative forcing of Earth's surface and a modest cooling of the lower stratosphere by approximately 0.5 °C (0.9 °F) per decade since the 1970s.

Natural Influences on Climate

There are a number of natural factors that influence Earth's climate. These factors include external influences such as explosive volcanic eruptions, natural variations in the output of the Sun, and slow changes in the configuration of Earth's orbit relative to the Sun. In addition, there are natural oscillations in Earth's climate that alter global patterns of wind circulation, precipitation, and surface temperatures. One such phenomenon is the El Niño/Southern Oscillation (ENSO), a coupled atmospheric and oceanic event that occurs in the Pacific Ocean every three to seven years. In addition, the Atlantic Multidecadal Oscillation (AMO) is a similar phenomenon that occurs over decades in the North Atlantic Ocean. Other types of oscillatory behaviour that produce dramatic shifts in climate may occur across timescales of centuries and millennia.

Volcanic Aerosols

Explosive volcanic eruptions have the potential to inject substantial amounts of sulfate aerosols into the lower stratosphere. In contrast to aerosol emissions in the lower troposphere, aerosols that enter the stratosphere may remain for several years before settling out, because of the relative absence of turbulent motions there. Consequently, aerosols from explosive volcanic eruptions have the potential to affect Earth's climate. Less-explosive eruptions, or eruptions that are less vertical in orientation, have a lower potential for substantial climate impact. Furthermore, because of large-scale circulation patterns within the stratosphere, aerosols injected within tropical regions tend to spread out over the globe, whereas aerosols injected within midlatitude and polar regions tend to remain confined to the middle and high latitudes of that hemisphere. Tropical eruptions, therefore, tend to have a greater climatic impact than eruptions occurring toward the poles. In 1991 the moderate eruption of Mount Pinatubo in the Philippines provided a peak forcing of approximately −4 watts per square metre and cooled the climate by about 0.5 °C (0.9 °F) over the following few years. By comparison, the 1815 Mount Tambora eruption in present-day Indonesia, typically implicated for the 1816 "year without a summer" in Europe and North America, is believed to have been associated with a radiative forcing of approximately −6 watts per square metre.

A column of gas and ash rising from Mount Pinatubo in the Philippines just days before the volcano's climactic explosion.

While in the stratosphere, volcanic sulfate aerosol actually absorbs longwave radiation emitted by Earth's surface, and absorption in the stratosphere tends to result in a cooling of the troposphere below. This vertical pattern of temperature change in the atmosphere influences the behaviour of winds in the lower atmosphere, primarily in winter. Thus, while there is essentially a global cooling effect for the first few years following an explosive volcanic eruption, changes in the winter patterns of surface winds may actually lead to warmer winters in some areas, such as Europe. Some modern examples of major eruptions include Krakatoa (Indonesia) in 1883, El Chichón (Mexico) in 1982, and Mount Pinatubo in 1991. There is also evidence that volcanic eruptions may influence other climate phenomena such as ENSO.

Variations in Solar Output

Direct measurements of solar irradiance, or solar output, have been available from satellites only since the late 1970s. These measurements show a very small peak-to-peak variation in solar irradiance (roughly 0.1 percent of the 1,366 watts per square metre received at the top of the atmosphere, for approximately 1.4 watts per square metre). However, indirect measures of solar activity are available from historical sunspot measurements dating back through the early 17th century. Attempts have been made to reconstruct graphs of solar irradiance variations from historical sunspot data by calibrating them against the measurements from modern satellites. However, since the modern measurements span only a few of the most recent 11-year solar cycles, estimates of solar output variability on 100-year and longer timescales are poorly correlated. Different assumptions regarding the relationship between the amplitudes of 11-year solar cycles and long-period solar output changes can lead to considerable differences in the resulting solar reconstructions. These differences in turn lead to fairly large uncertainty in estimating positive forcing by changes in solar irradiance since 1750. (Estimates range from 0.06 to 0.3 watt per square metre.) Even more challenging, given the lack of any modern analog, is the estimation of solar irradiance during the so-called Maunder Minimum, a period lasting from the mid-17th century to the early 18th century when very few sunspots were observed. While it is likely that solar irradiance was reduced at this time, it is difficult to calculate by how much. However, additional proxies of solar output exist that match reasonably well with the sunspot-derived records following the Maunder Minimum; these may be used as crude estimates of the solar irradiance variations.

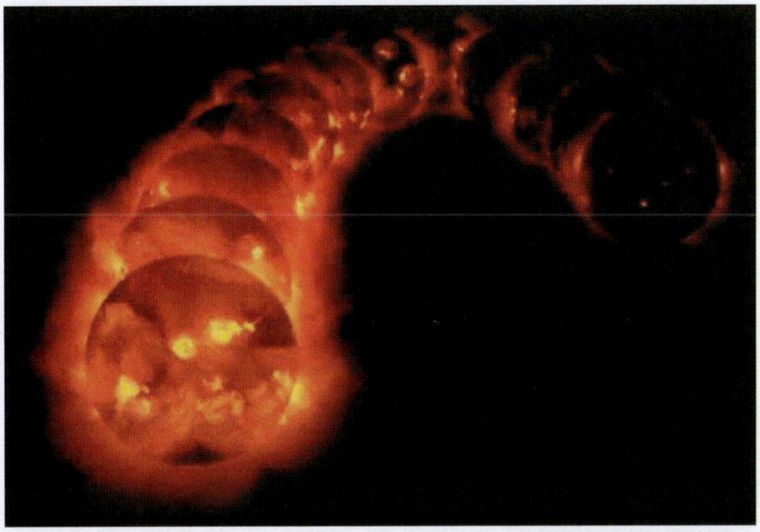

Twelve solar X-ray images obtained by Yohkoh. The solar coronal brightness decreases by a factor of about 100 during a solar cycle as the Sun goes from an "active" state (left) to a less active state (right).

In theory it is possible to estimate solar irradiance even farther back in time, over at least the past millennium, by measuring levels of cosmogenic isotopes such as carbon-14 and beryllium-10. Cosmogenic isotopes are isotopes that are formed by interactions of cosmic rays with atomic nuclei in the atmosphere and that subsequently fall to Earth, where they can be measured in the annual layers found in ice cores. Since their production rate in the upper atmosphere is modulated by changes in solar activity, cosmogenic isotopes may be used as indirect indicators of solar irradiance. However, as with the sunspot data, there is still considerable uncertainty in the amplitude of past solar variability implied by these data.

Solar forcing also affects the photochemical reactions that manufacture ozone in the stratos phere. Through this modulation of stratospheric ozone concentrations, changes in solar irradiance (particularly in the ultraviolet portion of the electromagnetic spectrum) can modify how both shortwave and longwave radiation in the lower stratosphere are absorbed. As a result, the vertical temperature profile of the atmosphere can change, and this change can in turn influence phenomena such as the strength of the winter jet streams.

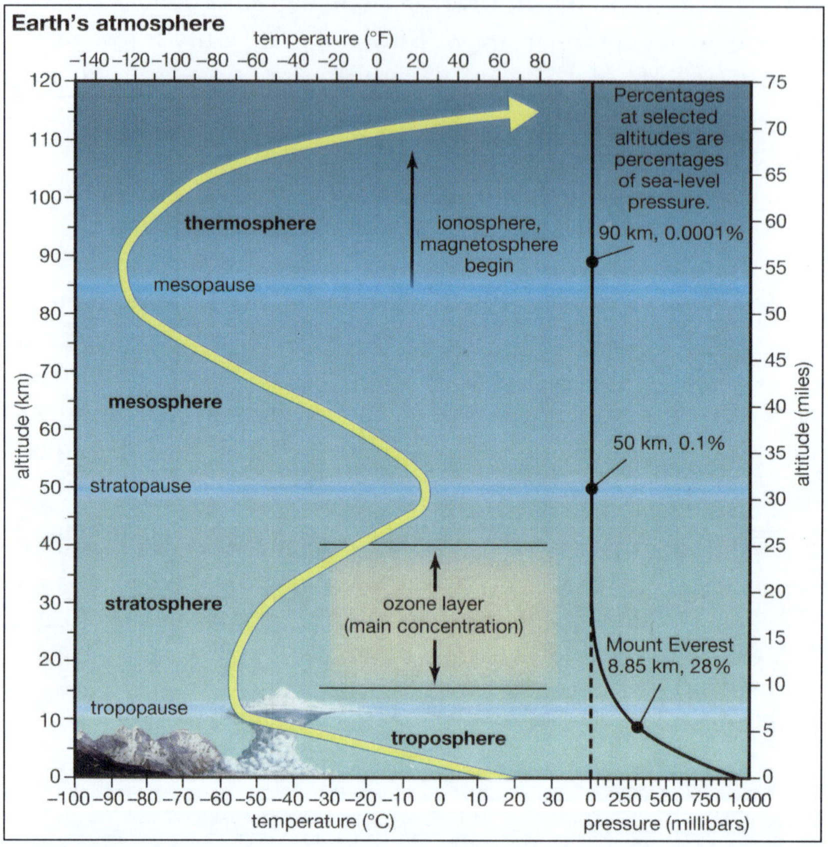

The layers of Earth's atmosphere. The yellow line shows the response of air temperature to increasing height.

Variations in Earth's Orbit

On timescales of tens of millennia, the dominant radiative forcing of Earth's climate is associated with slow variations in the geometry of Earth's orbit about the Sun. These variations include the precession of the equinoxes (that is, changes in the timing of summer and winter), occurring on a roughly 26,000-year timescale; changes in the tilt angle of Earth's rotational axis relative to the plane of Earth's orbit around the Sun, occurring on a roughly 41,000-year timescale; and changes in the eccentricity (the departure from a perfect circle) of Earth's orbit around the Sun, occurring on a roughly 100,000-year timescale. Changes in eccentricity slightly influence the mean annual solar radiation at the top of Earth's atmosphere, but the primary influence of all the orbital variations listed above is on the seasonal and latitudinal distribution of incoming solar radiation over Earth's surface. The major ice ages of the Pleistocene Epoch were closely related to the influence of these variations on summer insolation at high northern latitudes. Orbital variations thus exerted a primary control on the extent of continental ice sheets. However, Earth's orbital changes are generally believed to have had little impact on climate over the past few millennia, and so they are not considered to be significant factors in present-day climate variability.

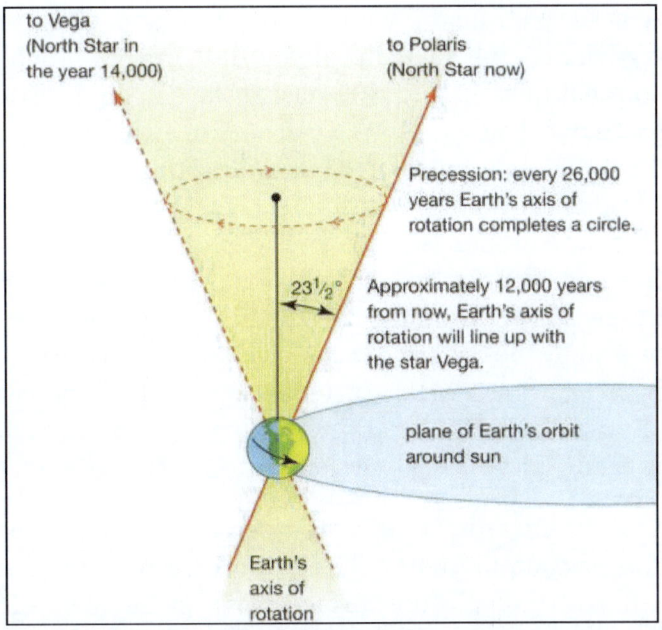

Earth's axis of rotation itself rotates, or precesses, completing one circle every 26,000 years. Consequently, Earth's North Pole points toward different stars (and sometimes toward empty space) as it travels in this circle. This precession is so slow that it is not noticeable in a person's lifetime, though astronomers must consider its effect when studying ancient sites such as Stonehenge.

Feedback Mechanisms and Climate Sensitivity

There are a number of feedback processes important to Earth's climate system and, in particular, its response to external radiative forcing. The most fundamental of these feedback mechanisms involves the loss of longwave radiation to space from the surface. Since this radiative loss increases with increasing surface temperatures according to the Stefan-Boltzmann law, it represents a stabilizing factor (that is, a negative feedback) with respect to near-surface air temperature.

Climate sensitivity can be defined as the amount of surface warming resulting from each additional watt per square metre of radiative forcing. Alternatively, it is sometimes defined as the warming that would result from a doubling of CO_2 concentrations and the associated addition of 4 watts per square metre of radiative forcing. In the absence of any additional feedbacks, climate sensitivity would be approximately 0.25 °C (0.45 °F) for each additional watt per square metre of radiative forcing. Stated alternatively, if the CO_2 concentration of the atmosphere present at the start of the industrial age (280 ppm) were doubled (to 560 ppm), the resulting additional 4 watts per square metre of radiative forcing would translate into a 1 °C (1.8 °F) increase in air temperature. However, there are additional feedbacks that exert a destabilizing, rather than stabilizing, influence, and these feedbacks tend to increase the sensitivity of climate to somewhere between 0.5 and 1.0 °C (0.9 and 1.8 °F) for each additional watt per square metre of radiative forcing.

Water Vapour Feedback

Unlike concentrations of other greenhouse gases, the concentration of water vapour in the atmosphere cannot freely vary. Instead, it is determined by the temperature of the lower atmosphere and surface through a physical relationship known as the Clausius-Clapeyron equation, named for 19th-century German physicist Rudolf Clausius and 19th-century French engineer Émile Clapeyron. Under the assumption that there is a liquid water surface in equilibrium with the atmosphere, this relationship indicates that an increase in the capacity of air to hold water vapour is a function of increasing temperature of that volume of air. This assumption is relatively good over the oceans, where water is plentiful, but not over the continents. For this reason the relative humidity (the percent of water vapour the air contains relative to its capacity) is approximately 100 percent over ocean regions and much lower over continental regions (approaching 0 percent in arid regions). Not surprisingly, the average relative humidity of Earth's lower atmosphere is similar to the fraction of Earth's surface covered by the oceans (that is, roughly 70 percent). This quantity is expected to remain approximately constant as Earth warms or cools. Slight changes to global relative humidity may result from human land-use modification, such as tropical deforestation and irrigation, which can affect the relative humidity over land areas up to regional scales.

The amount of water vapour in the atmosphere will rise as the temperature of the atmosphere rises. Since water vapour is a very potent greenhouse gas, even more potent than CO_2, the net greenhouse effect actually becomes stronger as the surface warms, which leads to even greater warming. This positive feedback is known as the "water vapour feedback". It is the primary reason that climate sensitivity is substantially greater than the previously stated theoretical value of 0.25 °C (0.45 °F) for each increase of 1 watt per square metre of radiative forcing.

Cloud Feedbacks

Different types of clouds form at different heights.

It is generally believed that as Earth's surface warms and the atmosphere's water vapour content increases, global cloud cover increases. However, the effects on near-surface air temperatures are complicated. In the case of low clouds, such as marine stratus clouds, the dominant radiative feature of the cloud is its albedo. Here any increase in low cloud cover acts in much the same way as an increase in surface ice cover: more incoming solar radiation is reflected and Earth's surface cools. On the other hand, high clouds, such as the towering cumulus clouds that extend up to the boundary between the troposphere and stratosphere, have a quite different impact on the surface radiation balance. The tops of cumulus clouds are considerably higher in the atmosphere and colder than their undersides. Cumulus cloud tops emit less longwave radiation out to space than the warmer cloud bottoms emit downward toward the surface. The end result of the formation of high cumulus clouds is greater warming at the surface.

The net feedback of clouds on rising surface temperatures is therefore somewhat uncertain. It represents a competition between the impacts of high and low clouds, and the balance is difficult to determine. Nonetheless, most estimates indicate that clouds on the whole represent a positive feedback and thus additional warming.

Ice Albedo Feedback

Another important positive climate feedback is the so-called ice albedo feedback. This feedback arises from the simple fact that ice is more reflective (that is, has a higher albedo) than land or water surfaces. Therefore, as global ice cover decreases, the reflectivity of Earth's surface decreases, more incoming solar radiation is absorbed by the surface, and the surface warms. This feedback is considerably more important when there is relatively extensive global ice cover, such as during the height of the last ice age, roughly 25,000 years ago. On a global scale the importance of ice albedo feedback decreases as Earth's surface warms and there is relatively less ice available to be melted.

Carbon Cycle Feedbacks

Another important set of climate feedbacks involves the global carbon cycle. In particular, the two main reservoirs of carbon in the climate system are the oceans and the terrestrial biosphere. These reservoirs have historically taken up large amounts of anthropogenic CO_2 emissions. Roughly 50–70 percent is removed by the oceans, whereas the remainder is taken up by the terrestrial biosphere. Global warming, however, could decrease the capacity of these reservoirs to sequester atmospheric CO_2. Reductions in the rate of carbon uptake by these reservoirs would increase the pace of CO_2 buildup in the atmosphere and represent yet another possible positive feedback to increased greenhouse gas concentrations.

In the world's oceans, this feedback effect might take several paths. First, as surface waters warm, they would hold less dissolved CO_2. Second, if more CO_2 were added to the atmosphere and taken up by the oceans, bicarbonate ions (HCO_3^-) would multiply and ocean acidity would increase. Since calcium carbonate ($CaCO_3$) is broken down by acidic solutions, rising acidity would threaten ocean-dwelling fauna that incorporate $CaCO_3$ into their skeletons or shells. As it becomes increasingly difficult for these organisms to absorb oceanic carbon, there would be a corresponding decrease in the efficiency of the biological pump that helps to maintain the oceans as a carbon sink. Third, rising surface temperatures might lead to a slowdown in the so-called thermohaline circulation, a global pattern of oceanic flow that partly drives the sinking of surface waters near the poles and is responsible for much of the burial of carbon in the deep ocean. A slowdown in this flow due to an influx of melting fresh water into what are normally saltwater conditions might also cause the solubility pump, which transfers CO_2 from shallow to deeper waters, to become less efficient. Indeed, it is predicted that if global warming continued to a certain point, the oceans would cease to be a net sink of CO_2 and would become a net source.

As large sections of tropical forest are lost because of the warming and drying of regions such as Amazonia, the overall capacity of plants to sequester atmospheric CO_2 would be reduced. As a result, the terrestrial biosphere, though currently a carbon sink, would become a carbon source. Ambient temperature is a significant factor affecting the pace of photosynthesis in plants, and many plant species that are well adapted to their local climatic conditions have maximized their photosynthetic rates. As temperatures increase and conditions begin to exceed the optimal temperature range for both photosynthesis and soil respiration, the rate of photosynthesis would decline. As dead plants decompose, microbial metabolic activity (a CO_2 source) would increase and would eventually outpace photosynthesis.

Under sufficient global warming conditions, methane sinks in the oceans and terrestrial biosphere also might become methane sources. Annual emissions of methane by wetlands might either increase or decrease, depending on temperatures and input of nutrients, and it is possible that wetlands could switch from source to sink. There is also the potential for increased methane release as a result of the warming of Arctic permafrost (on land) and further methane release at the continental margins of the oceans (a few hundred metres below sea level). The current average atmospheric methane concentration of 1,750 ppb is equivalent to 3.5 gigatons (3.5 billion tons) of carbon. There are at least 400 gigatons of carbon equivalent stored in Arctic permafrost and as much as 10,000 gigatons (10 trillion tons) of carbon equivalent trapped on the continental margins of the oceans in a hydrated crystalline form known as clathrate. It is believed that some fraction of this trapped methane could become unstable with additional warming, although the amount and rate of potential emission remain highly uncertain.

GREENHOUSE EFFECT

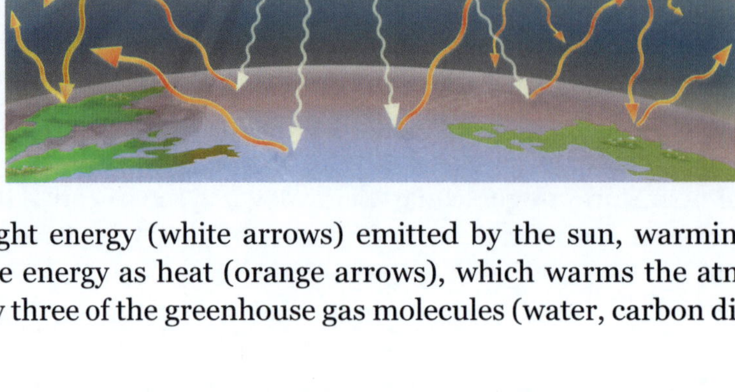

Diagram showing light energy (white arrows) emitted by the sun, warming the earth's surface which then emits the energy as heat (orange arrows), which warms the atmosphere and is then re-emitted as heat by three of the greenhouse gas molecules (water, carbon dioxide, and methane).

The greenhouse effect is the process by which radiation from a planet's atmosphere warms the planet's surface to a temperature above what it would be without this atmosphere.

Radiatively active gases (i.e., greenhouse gases) in a planet's atmosphere radiate energy in all directions. Part of this radiation is directed towards the surface, warming it. The intensity of the downward radiation – that is, the strength of the greenhouse effect – will depend on the atmosphere's temperature and on the amount of greenhouse gases that the atmosphere contains.

Earth's natural greenhouse effect is critical to supporting life, and initially was a precursor to life moving out of the ocean onto land. Human activities, however, mainly the burning of fossil fuels and clearcutting of forests, have accelerated the greenhouse effect and caused global warming.

The planet Venus experienced runaway greenhouse effect, resulting in an atmosphere which is 96% carbon dioxide, with surface atmospheric pressure roughly the same as found 900 m (3,000 ft) underwater on Earth. Venus may have had water oceans, but they would have boiled off as the mean surface temperature rose to 735 K (462 °C; 863 °F).

The term "greenhouse effect" continues to see use in scientific circles and the media despite being a slight misnomer, as an atmosphere reduces radiative heat loss while a greenhouse blocks convective heat loss. The result, however, is an increase in temperature in both cases.

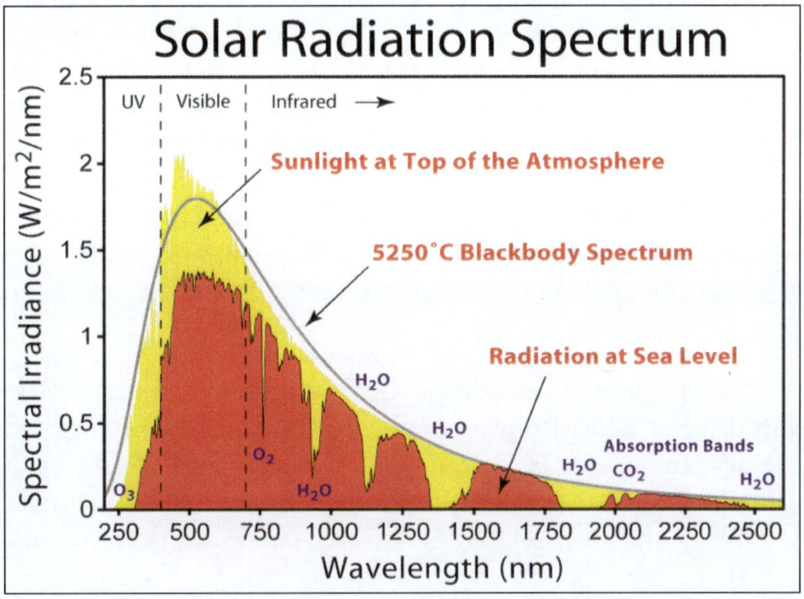

The solar radiation spectrum for direct light at both the top of Earth's atmosphere and at sea level.

Earth receives energy from the Sun in the form of ultraviolet, visible, and near-infrared radiation. About 26% of the incoming solar energy is reflected to space by the atmosphere and clouds, and 19% is absorbed by the atmosphere and clouds. Most of the remaining energy is absorbed at the surface of Earth. Because the Earth's surface is colder than the Sun, it radiates at wavelengths that are much longer than the wavelengths that were absorbed. Most of this thermal radiation is absorbed by the atmosphere and warms it. The atmosphere also gains heat by sensible and latent heat fluxes from the surface. The atmosphere radiates energy both upwards and downwards; the part radiated downwards is absorbed by the surface of Earth. This leads to a higher equilibrium temperature than if the atmosphere did not radiate.

An ideal thermally conductive blackbody at the same distance from the Sun as Earth would have a temperature of about 5.3 °C (41.5 °F). However, because Earth reflects about 30% of the incoming sunlight, this idealized planet's effective temperature (the temperature of a blackbody that would emit the same amount of radiation) would be about −18 °C (0 °F). The surface temperature of this hypothetical planet is 33 °C (59 °F) below Earth's actual surface temperature of approximately 14 °C (57 °F). The greenhouse effect is the contribution of greenhouse gases to this difference.

Mechanism

The basic mechanism can be qualified in a number of ways, none of which affect the fundamental process. The atmosphere near the surface is largely opaque to thermal radiation (with important exceptions for "window" bands), and most heat loss from the surface is by sensible heat and latent heat transport. Radiative energy losses become increasingly important higher in the atmosphere, largely because of the decreasing concentration of water vapor, an important greenhouse gas. It is more realistic to think of the greenhouse effect as applying to a layer in the mid-troposphere, which is effectively coupled to the surface by a lapse rate. The simple picture also assumes a steady state, but in the real world, the diurnal cycle as well as the seasonal cycle and weather disturbances complicate matters. Solar heating applies only during daytime. During the night, the atmosphere cools somewhat, but not greatly, because its emissivity is low. Diurnal temperature changes decrease with height in the atmosphere.

Within the region where radiative effects are important, the description given by the idealized greenhouse model becomes realistic. Earth's surface, warmed to a temperature around 255 K, radiates long-wavelength, infrared heat in the range of 4–100 μm. At these wavelengths, greenhouse gases that were largely transparent to incoming solar radiation are more absorbent. Each layer of atmosphere with greenhouses gases absorbs some of the heat being radiated upwards from lower layers. It reradiates in all directions, both upwards and downwards; in equilibrium (by definition) the same amount as it has absorbed. This results in more warmth below. Increasing the concentration of the gases increases the amount of absorption and reradiation, and thereby further warms the layers and ultimately the surface below.

Greenhouse gases—including most diatomic gases with two different atoms (such as carbon monoxide, CO) and all gases with three or more atoms—are able to absorb and emit infrared radiation. Though more than 99% of the dry atmosphere is IR transparent (because the main constituents—N_2, O_2, and Ar—are not able to directly absorb or emit infrared radiation), intermolecular collisions cause the energy absorbed and emitted by the greenhouse gases to be shared with the other, non-IR-active, gases.

Greenhouse Gases

By their percentage contribution to the greenhouse effect on Earth the four major gases are:

Atmospheric gases only absorb some wavelengths of energy but are transparent to others. The absorption patterns of water vapor (blue peaks) and carbon dioxide (pink peaks) overlap in some wavelengths. Carbon dioxide is not as strong a greenhouse gas as water vapor, but it absorbs energy in longer wavelengths (12–15 micrometers) that water vapor does not, partially closing the "window" through which heat radiated by the surface would normally escape to space:

- Water vapor, 36–70%.

- Carbon dioxide, 9–26%.
- Methane, 4–9%.
- Ozone, 3–7%.

It is not possible to assign a specific percentage to each gas because the absorption and emission bands of the gases overlap (hence the ranges given above). Clouds also absorb and emit infrared radiation and thus affect the radiative properties of the atmosphere.

Role in Climate Change

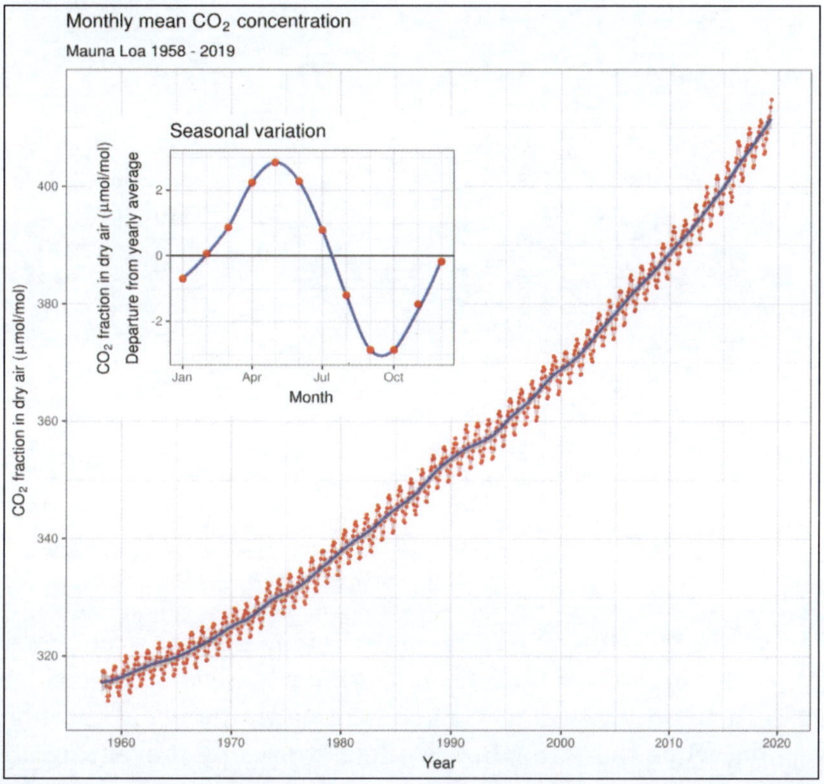

The Keeling Curve of atmospheric CO_2 concentrations measured at Mauna Loa Observatory.

Strengthening of the greenhouse effect through human activities is known as the enhanced (or anthropogenic) greenhouse effect. This increase in radiative forcing from human activity is attributable mainly to increased atmospheric carbon dioxide levels. According to the latest Assessment Report from the Intergovernmental Panel on Climate Change, "atmospheric concentrations of carbon dioxide, methane and nitrous oxide are unprecedented in at least the last 800,000 years. Their effects, together with those of other anthropogenic drivers, have been detected throughout the climate system and are extremely likely to have been the dominant cause of the observed warming since the mid-20th century".

CO_2 is produced by fossil fuel burning and other activities such as cement production and tropical deforestation. Measurements of CO_2 from the Mauna Loa observatory show that concentrations have increased from about 313 parts per million (ppm) in 1960 to about 389 ppm in 2010. It reached the 400 ppm milestone on May 9, 2013. The current observed amount of CO_2 exceeds the geological record maxima (~300 ppm) from ice core data. The effect of combustion-produced carbon dioxide on the global climate, a special case of the greenhouse effect first described in 1896 by Svante Arrhenius, has also been called the Callendar effect.

Over the past 800,000 years, ice core data shows that carbon dioxide has varied from values as low as 180 ppm to the pre-industrial level of 270 ppm. Paleoclimatologists consider variations in carbon dioxide concentration to be a fundamental factor influencing climate variations over this time scale.

Real Greenhouses

A modern Greenhouse in RHS Wisley.

The "greenhouse effect" of the atmosphere is named by analogy to greenhouses which become warmer in sunlight. However, a greenhouse is not primarily warmed by the "greenhouse effect". "Greenhouse effect" is actually a misnomer since heating in the usual greenhouse is due to the reduction of convection, while the "greenhouse effect" works by preventing absorbed heat from leaving the structure through radiative transfer.

A greenhouse is built of any material that passes sunlight: usually glass or plastic. The sun warms the ground and contents inside just like the outside, and these then warm the air. Outside, the warm air near the surface rises and mixes with cooler air aloft, keeping the temperature lower than inside, where the air continues to heat up because it is confined within the greenhouse. This can be demonstrated by opening a small window near the roof of a greenhouse: the temperature will drop considerably. It was demonstrated experimentally that a (not heated) "greenhouse" with a cover of rock salt (which is transparent to infrared) heats up an enclosure similarly to one with a glass cover. Thus greenhouses work primarily by preventing convective cooling.

Heated greenhouses are yet another matter: as they have an internal source of heating, it is desirable to minimise the amount of heat leaking out by radiative cooling. This can be done through the use of adequate glazing.

Related Effects

Anti-Greenhouse Effect

The anti-greenhouse effect is a mechanism similar and symmetrical to the greenhouse effect: in the greenhouse effect, the atmosphere lets radiation in while not letting thermal radiation out, thus warming the body surface; in the anti-greenhouse effect, the atmosphere keeps radiation out while letting thermal radiation out, which lowers the equilibrium surface temperature. Such an effect has been proposed for Saturn's moon Titan.

Runaway Greenhouse Effect

A runaway greenhouse effect occurs if positive feedbacks lead to the evaporation of all greenhouse gases into the atmosphere. A runaway greenhouse effect involving carbon dioxide and water vapor has long ago been hypothesized to have occurred on Venus, this idea is still largely accepted.

Greenhouse Gas

The greenhouse effect of solar radiation on the Earth's surface caused by greenhouse gases.

A greenhouse gas (sometimes abbreviated GHG) is a gas that absorbs and emits radiant energy within the thermal infrared range. Greenhouse gases cause the greenhouse effect. The primary greenhouse gases in Earth's atmosphere are water vapor, carbon dioxide, methane, nitrous oxide and ozone. Without greenhouse gases, the average temperature of Earth's surface would be about −18 °C (0 °F), rather than the present average of 15 °C (59 °F). The atmospheres of Venus, Mars and Titan also contain greenhouse gases.

Human activities since the beginning of the Industrial Revolution have produced a 45% increase in the atmospheric concentration of carbon dioxide (CO_2), from 280 ppm in 1750 to 415 ppm in 2019. This increase has occurred despite the uptake of more than half of the emissions by various natural "sinks" involved in the carbon cycle. The vast majority of anthropogenic carbon dioxide emissions come from combustion of fossil fuels, principally coal, oil, and natural gas, with additional contributions coming from deforestation, changes in land use, soil erosion and agriculture (including livestock).

Should greenhouse gas emissions continue at their rate in 2019, global warming could cause Earth's surface temperature to exceed historical values as early as 2047, with potentially harmful effects on ecosystems, biodiversity and human livelihoods. At current emission rates, temperatures could increase by 2 °C, which the United Nations' IPCC designated as the upper limit to avoid "dangerous" levels, by 2036.

Gases in Earth's Atmosphere

The most common gases in Earth's atmosphere are nitrogen (78%), oxygen (21%), and argon (0.9%). The next most common gases are carbon dioxide (0.04%), nitrous oxide, methane, and ozone. They are trace gases that account for almost one tenth of 1% of Earth's atmosphere.

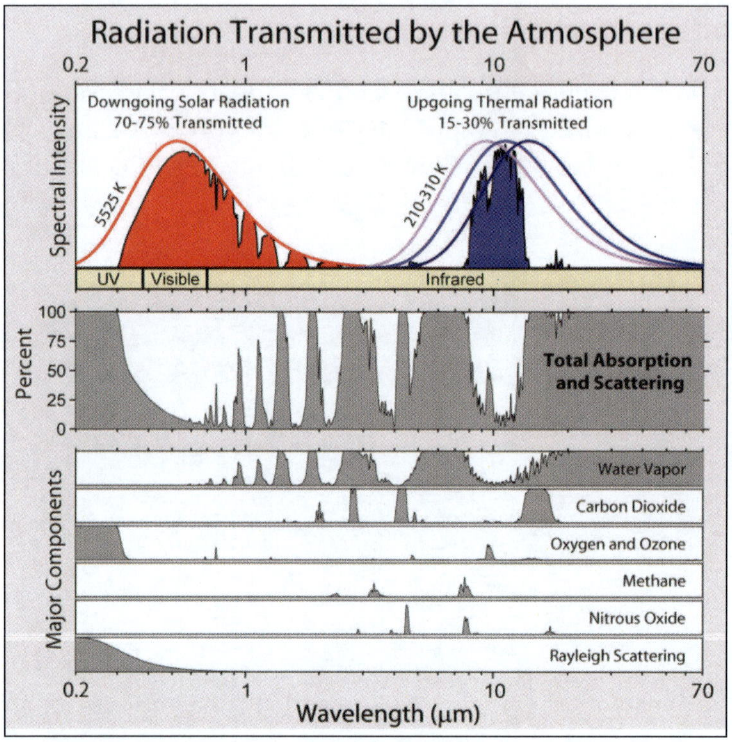

Atmospheric absorption and scattering at different wavelengths of electromagnetic waves. The largest absorption band of carbon dioxide is not far from the maximum in the thermal emission from ground, and it partly closes the window of transparency of water; hence its major effect.

Greenhouse Gases

Greenhouse gases are those that absorb and emit infrared radiation in the wavelength range emitted by Earth. In order, the most abundant greenhouse gases in Earth's atmosphere are:

- Water vapor (H_2O).
- Carbon dioxide (CO_2).
- Methane (CH_4).
- Nitrous oxide (N_2O).
- Ozone (O_3).

- Chlorofluorocarbons (CFCs).
- Hydrofluorocarbons (includes HCFCs and HFCs).

Atmospheric concentrations are determined by the balance between sources (emissions of the gas from human activities and natural systems) and sinks (the removal of the gas from the atmosphere by conversion to a different chemical compound or absorption by bodies of water). The proportion of an emission remaining in the atmosphere after a specified time is the "airborne fraction" (AF). The *annual airborne fraction* is the ratio of the atmospheric increase in a given year to that year's total emissions. As of 2006 the annual airborne fraction for CO_2 was about 0.45. The annual airborne fraction increased at a rate of 0.25 ± 0.21% per year over the period 1959–2006.

Non-greenhouse Gases

The major atmospheric constituents: nitrogen (N_2), oxygen (O_2), and argon (Ar) are not greenhouse gases because molecules containing two atoms of the same element such as N_2 and O_2 have no net change in the distribution of their electrical charges when they vibrate, and monatomic gases such as Ar do not have vibrational modes. Hence they are almost totally unaffected by infrared radiation. Some molecules containing just two atoms of different elements, such as carbon monoxide (CO) and hydrogen chloride (HCl), do absorb infrared radiation, but these molecules are short-lived in the atmosphere owing to their reactivity and solubility. Therefore, they do not contribute significantly to the greenhouse effect and often are omitted when discussing greenhouse gases.

Indirect Radiative Effects

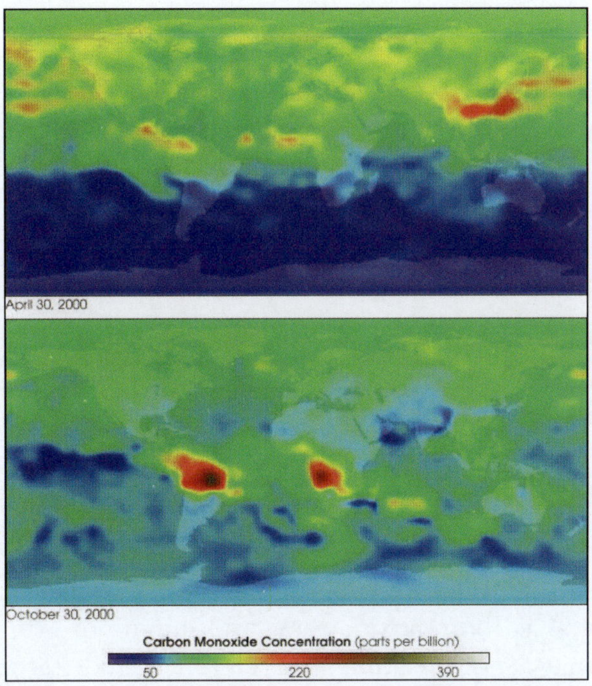

The false colors in this image represent concentrations of carbon monoxide in the lower atmosphere, ranging from about 390 parts per billion (dark brown pixels), to 220 parts per billion (red pixels), to 50 parts per billion (blue pixels).

Some gases have indirect radiative effects (whether or not they are greenhouse gases themselves). This happens in two main ways. One way is that when they break down in the atmosphere they produce another greenhouse gas. For example, methane and carbon monoxide (CO) are oxidized to give carbon dioxide (and methane oxidation also produces water vapor). Oxidation of CO to CO_2 directly produces an unambiguous increase in radiative forcing although the reason is subtle. The peak of the thermal IR emission from Earth's surface is very close to a strong vibrational absorption band of CO_2 (wavelength 15 microns, or wavenumber 667 cm^{-1}). On the other hand, the single CO vibrational band only absorbs IR at much shorter wavelengths (4.7 microns, or 2145 cm^{-1}), where the emission of radiant energy from Earth's surface is at least a factor of ten lower. Oxidation of methane to CO_2, which requires reactions with the OH radical, produces an instantaneous reduction in radiative absorption and emission since CO_2 is a weaker greenhouse gas than methane. However, the oxidations of CO and CH_4 are entwined since both consume OH radicals. In any case, the calculation of the total radiative effect includes both direct and indirect forcing.

A second type of indirect effect happens when chemical reactions in the atmosphere involving these gases change the concentrations of greenhouse gases. For example, the destruction of non-methane volatile organic compounds (NMVOCs) in the atmosphere can produce ozone. The size of the indirect effect can depend strongly on where and when the gas is emitted.

Methane has indirect effects in addition to forming CO_2. The main chemical that reacts with methane in the atmosphere is the hydroxyl radical (OH), thus more methane means that the concentration of OH goes down. Effectively, methane increases its own atmospheric lifetime and therefore its overall radiative effect. The oxidation of methane can produce both ozone and water; and is a major source of water vapor in the normally dry stratosphere. CO and NMVOCs produce CO_2 when they are oxidized. They remove OH from the atmosphere, and this leads to higher concentrations of methane. The surprising effect of this is that the global warming potential of CO is three times that of CO_2. The same process that converts NMVOCs to carbon dioxide can also lead to the formation of tropospheric ozone. Halocarbons have an indirect effect because they destroy stratospheric ozone. Finally, hydrogen can lead to ozone production and CH_4 increases as well as producing stratospheric water vapor.

Contribution of Clouds to Earth's Greenhouse Effect

The major non-gas contributor to Earth's greenhouse effect, clouds, also absorb and emit infrared radiation and thus have an effect on greenhouse gas radiative properties. Clouds are water droplets or ice crystals suspended in the atmosphere.

Impacts on the Overall Greenhouse Effect

Schmidt analysed how individual components of the atmosphere contribute to the total greenhouse effect. They estimated that water vapor accounts for about 50% of Earth's greenhouse effect, with clouds contributing 25%, carbon dioxide 20%, and the minor greenhouse gases and aerosols accounting for the remaining 5%. In the study, the reference model atmosphere is for 1980 conditions.

The contribution of each gas to the greenhouse effect is determined by the characteristics of that gas, its abundance, and any indirect effects it may cause. For example, the direct radiative effect of a mass of methane is about 84 times stronger than the same mass of carbon dioxide over a 20-year time frame but it is present in much smaller concentrations so that its total direct radiative effect is smaller, in part due to its shorter atmospheric lifetime. On the other hand, in addition to its direct radiative impact, methane has a large, indirect radiative effect because it contributes to ozone formation. the contribution to climate change from methane is at least double previous estimates as a result of this effect.

When ranked by their direct contribution to the greenhouse effect, the most important are:

Compound	Formula	Concentration in atmosphere (ppm)	Contribution (%)
Water vapor and clouds	H_2O	10–50,000[(A)]	36–72%
Carbon dioxide	CO_2	~400	9–26%
Methane	CH_4	~1.8	4–9%
Ozone	O_3	2–8[(B)]	3–7%
(A) Water vapor strongly varies locally (B) The concentration in stratosphere. About 90% of the ozone in Earth's atmosphere is contained in the stratosphere.			

In addition to the main greenhouse gases listed above, other greenhouse gases include sulfur hexafluoride, hydrofluorocarbons and perfluorocarbons. Some greenhouse gases are not often listed. For example, nitrogen trifluoride has a high global warming potential (GWP) but is only present in very small quantities.

Proportion of Direct Effects at a Given Moment

It is not possible to state that a certain gas causes an exact percentage of the greenhouse effect. This is because some of the gases absorb and emit radiation at the same frequencies as others, so that the total greenhouse effect is not simply the sum of the influence of each gas. The higher ends of the ranges quoted are for each gas alone; the lower ends account for overlaps with the other gases. In addition, some gases, such as methane, are known to have large indirect effects that are still being quantified.

Atmospheric Lifetime

Aside from water vapor, which has a residence time of about nine days, major greenhouse gases are well mixed and take many years to leave the atmosphere. Although it is not easy to know with precision how long it takes greenhouse gases to leave the atmosphere, there are estimates for the principal greenhouse gases. Jacob defines the lifetime τ of an atmospheric species X in a one-box model as the average time that a molecule of X remains in the box. Mathematically τ can be defined as the ratio of the mass m (in kg) of X in the box to its removal rate, which is the sum of the flow of X out of the box (F_{out}), chemical loss of X (L), and deposition of X (D) (all in kg/s): $\tau = \frac{m}{F_{out} + L + D}$. If output of this gas into the box ceased, then after time τ, its concentration would decrease by about 63%.

The atmospheric lifetime of a species therefore measures the time required to restore equilibrium following a sudden increase or decrease in its concentration in the atmosphere. Individual atoms or molecules may be lost or deposited to sinks such as the soil, the oceans and other waters, or vegetation and other biological systems, reducing the excess to background concentrations. The average time taken to achieve this is the mean lifetime.

Carbon dioxide has a variable atmospheric lifetime, and cannot be specified precisely. The atmospheric lifetime of CO_2 is estimated of the order of 30–95 years. This figure accounts for CO_2 molecules being removed from the atmosphere by mixing into the ocean, photosynthesis, and other processes. However, this excludes the balancing fluxes of CO_2 into the atmosphere from the geological reservoirs, which have slower characteristic rates. Although more than half of the CO_2 emitted is removed from the atmosphere within a century, some fraction (about 20%) of emitted CO_2 remains in the atmosphere for many thousands of years. Similar issues apply to other greenhouse gases, many of which have longer mean lifetimes than CO_2, e.g. N_2O has a mean atmospheric lifetime of 121 years.

Radiative Forcing

Earth absorbs some of the radiant energy received from the sun, reflects some of it as light and reflects or radiates the rest back to space as heat. Earth's surface temperature depends on this balance between incoming and outgoing energy. If this energy balance is shifted, Earth's surface becomes warmer or cooler, leading to a variety of changes in global climate.

A number of natural and man-made mechanisms can affect the global energy balance and force changes in Earth's climate. Greenhouse gases are one such mechanism. Greenhouse gases absorb and emit some of the outgoing energy radiated from Earth's surface, causing that heat to be retained in the lower atmosphere. As explained above, some greenhouse gases remain in the atmosphere for decades or even centuries, and therefore can affect Earth's energy balance over a long period. Radiative forcing quantifies the effect of factors that influence Earth's energy balance, including changes in the concentrations of greenhouse gases. Positive radiative forcing leads to warming by increasing the net incoming energy, whereas negative radiative forcing leads to cooling.

Global Warming Potential

The global warming potential (GWP) depends on both the efficiency of the molecule as a greenhouse gas and its atmospheric lifetime. GWP is measured relative to the same mass of CO_2 and evaluated for a specific timescale. Thus, if a gas has a high (positive) radiative forcing but also a short lifetime, it will have a large GWP on a 20-year scale but a small one on a 100-year scale. Conversely, if a molecule has a longer atmospheric lifetime than CO_2 its GWP will increase when the timescale is considered. Carbon dioxide is defined to have a GWP of 1 over all time periods.

Methane has an atmospheric lifetime of 12 ± 3 years. The 2007 IPCC report lists the GWP as 72 over a time scale of 20 years, 25 over 100 years and 7.6 over 500 years. A 2014 analysis, however, states that although methane's initial impact is about 100 times greater than that of CO_2, because of the shorter atmospheric lifetime, after six or seven decades, the impact of the two gases is about equal, and from then on methane's relative role continues to decline. The decrease in GWP at longer times is because methane is degraded to water and CO_2 through chemical reactions in the atmosphere.

Examples of the atmospheric lifetime and GWP relative to CO_2 for several greenhouse gases are given in the following table:

Atmospheric lifetime and GWP relative to CO_2 at different time horizon for various greenhouse gases					
Gas name	Chemical formula	Lifetime (years)	Global warming potential (GWP) for given time horizon		
			20-yr	100-yr	500-yr
Carbon dioxide	CO_2	30–95	1	1	1
Methane	CH_4	12	84	28	7.6
Nitrous oxide	N_2O	121	264	265	153
CFC-12	CCl_2F_2	100	10 800	10 200	5 200
HCFC-22	$CHClF_2$	12	5 280	1 760	549
Tetrafluoromethane	CF_4	50 000	4 880	6 630	11 200
Hexafluoroethane	C_2F_6	10 000	8 210	11 100	18 200
Sulfur hexafluoride	SF_6	3 200	17 500	23 500	32 600
Nitrogen trifluoride	NF_3	500	12 800	16 100	20 700

The use of CFC-12 (except some essential uses) has been phased out due to its ozone depleting properties. The phasing-out of less active HCFC-compounds will be completed in 2030.

Carbon dioxide in Earth's atmosphere if half of global-warming emissions are not absorbed.

Natural and Anthropogenic Sources

This diagram shows a simplified representation of the contemporary global carbon cycle. Changes are measured in gigatons of carbon per year (GtC/y). Canadell estimated the growth rate of global average atmospheric CO_2 for 2000–2006 as 1.93 parts-per-million per year (4.1 petagrams of carbon per year).

Aside from purely human-produced synthetic halocarbons, most greenhouse gases have both natural and human-caused sources. During the pre-industrial Holocene, concentrations of existing gases were roughly constant, because the large natural sources and sinks roughly balanced. In the industrial era, human activities have added greenhouse gases to the atmosphere, mainly through the burning of fossil fuels and clearing of forests.

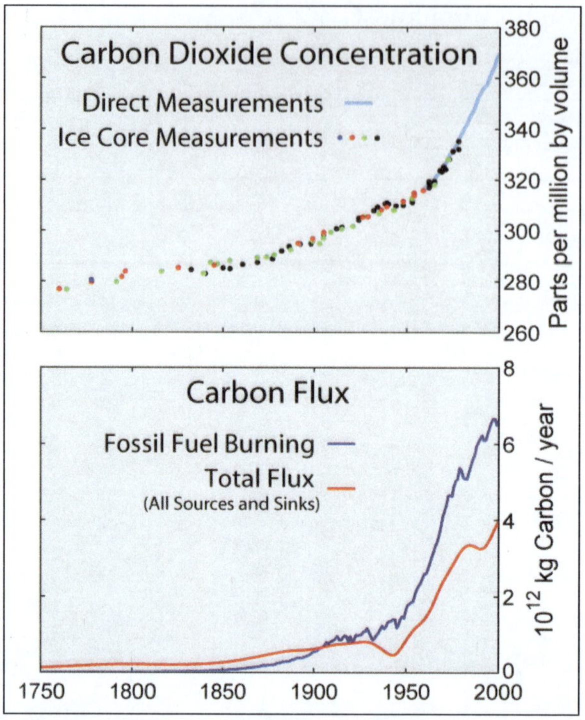

Top: Increasing atmospheric carbon dioxide levels as measured in the atmosphere and reflected in ice cores. Bottom: The amount of net carbon increase in the atmosphere, compared to carbon emissions from burning fossil fuel.

The 2007 Fourth Assessment Report compiled by the IPCC (AR4) noted that "changes in atmospheric concentrations of greenhouse gases and aerosols, land cover and solar radiation alter the energy balance of the climate system", and concluded that "increases in anthropogenic greenhouse gas concentrations is very likely to have caused most of the increases in global average temperatures since the mid-20th century". In AR4, "most of" is defined as more than 50%.

Current greenhouse gas concentrations					
Gas	Pre-1750 tropospheric concentration	Recent tropospheric concentration	Absolute increase since 1750	Percentage increase since 1750	Increased radiative forcing (W/m^2)
Carbon dioxide (CO_2)	280 ppm	395.4 ppm	115.4 ppm	41.2%	1.88
Methane (CH_4)	700 ppb	1893 ppb / 1762 ppb	1193 ppb / 1062 ppb	170.4% / 151.7%	0.49
Nitrous oxide (N_2O)	270 ppb	326 ppb / 324 ppb	56 ppb / 54 ppb	20.7% / 20.0%	0.17
Tropospheric ozone (O_3)	237 ppb	337 ppb	100 ppb	42%	0.4

Relevant to radiative forcing and ozone depletion; all of the following have no natural sources and hence zero amounts pre-industrial.		
Gas	Recent tropospheric concentration	Increased radiative forcing (W/m²)
CFC-11 (trichlorofluoromethane) (CCl_3F)	236 ppt / 234 ppt	0.061
CFC-12 (CCl_2F_2)	527 ppt / 527 ppt	0.169
CFC-113 (Cl_2FC-$CClF_2$)	74 ppt / 74 ppt	0.022
HCFC-22 ($CHClF_2$)	231 ppt / 210 ppt	0.046
HCFC-141b (CH_3CCl_2F)	24 ppt / 21 ppt	0.0036
HCFC-142b (CH_3CClF_2)	23 ppt / 21 ppt	0.0042
Halon 1211 ($CBrClF_2$)	4.1 ppt / 4.0 ppt	0.0012
Halon 1301 ($CBrClF_3$)	3.3 ppt / 3.3 ppt	0.001
HFC-134a (CH_2FCF_3)	75 ppt / 64 ppt	0.0108
Carbon tetrachloride (CCl_4)	85 ppt / 83 ppt	0.0143
Sulfur hexafluoride (SF_6)	7.79 ppt / 7.39 ppt	0.0043
Other halocarbons	Varies by substance	collectively 0.02
Halocarbons in total		0.3574

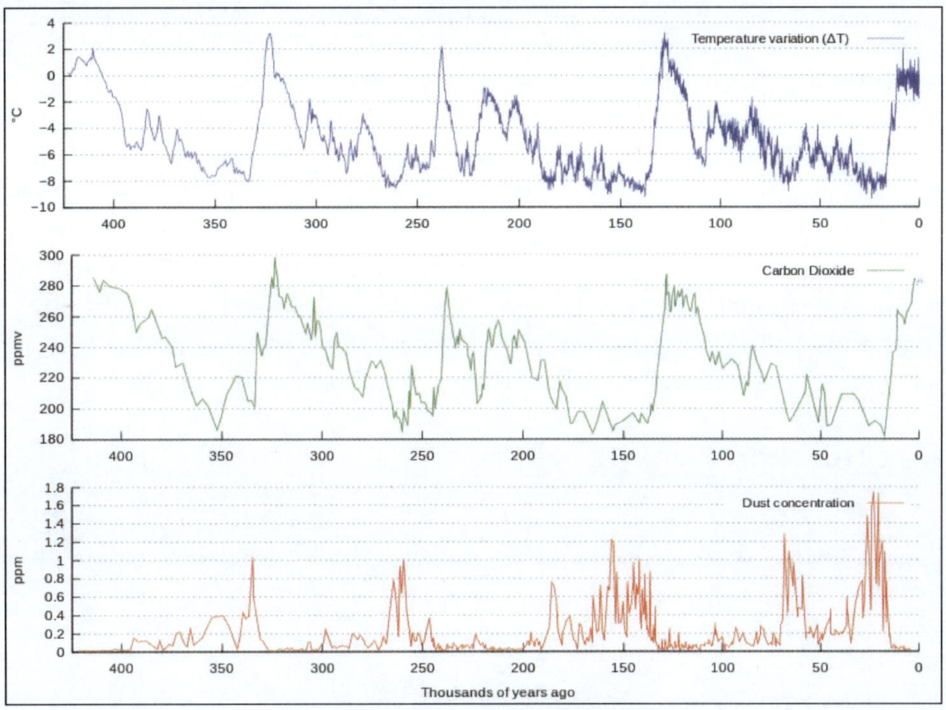

400,000 years of ice core data.

Ice cores provide evidence for greenhouse gas concentration variations over the past 800,000 years. Both CO_2 and CH_4 vary between glacial and interglacial phases, and concentrations of these gases correlate strongly with temperature. Direct data does not exist for periods earlier than those represented in the ice core record, a record that indicates CO_2 mole fractions stayed within a range of 180 ppm to 280 ppm throughout the last 800,000 years, until the increase of the last 250 years. However, various proxies and modeling suggests larger variations in past epochs; 500 million years ago CO_2 levels were likely 10 times higher than now. Indeed, higher CO_2 concentrations are thought to have prevailed throughout most of the Phanerozoic eon, with concentrations four to six times current concentrations during the Mesozoic era, and ten to fifteen times current concentrations during the early Palaeozoic era until the middle of the Devonian period, about 400 Ma. The spread of land plants is thought to have reduced CO_2 concentrations during the late Devonian, and plant activities as both sources and sinks of CO_2 have since been important in providing stabilising feedbacks. Earlier still, a 200-million year period of intermittent, widespread glaciation extending close to the equator (Snowball Earth) appears to have been ended suddenly, about 550 Ma, by a colossal volcanic outgassing that raised the CO_2 concentration of the atmosphere abruptly to 12%, about 350 times modern levels, causing extreme greenhouse conditions and carbonate deposition as limestone at the rate of about 1 mm per day. This episode marked the close of the Precambrian eon, and was succeeded by the generally warmer conditions of the Phanerozoic, during which multicellular animal and plant life evolved. No volcanic carbon dioxide emission of comparable scale has occurred since. In the modern era, emissions to the atmosphere from volcanoes are approximately 0.645 billion tonnes of CO_2 per year, whereas humans contribute 29 billion tonnes of CO_2 each year.

Ice Cores

Measurements from Antarctic ice cores show that before industrial emissions started atmospheric CO_2 mole fractions were about 280 parts per million (ppm), and stayed between 260 and 280 during the preceding ten thousand years. Carbon dioxide mole fractions in the atmosphere have gone up by approximately 35 percent since the 1900s, rising from 280 parts per million by volume to 387 parts per million in 2009. One study using evidence from stomata of fossilized leaves suggests greater variability, with carbon dioxide mole fractions above 300 ppm during the period seven to ten thousand years ago, though others have argued that these findings more likely reflect calibration or contamination problems rather than actual CO_2 variability. Because of the way air is trapped in ice (pores in the ice close off slowly to form bubbles deep within the firn) and the time period represented in each ice sample analyzed, these figures represent averages of atmospheric concentrations of up to a few centuries rather than annual or decadal levels.

Changes since the Industrial Revolution

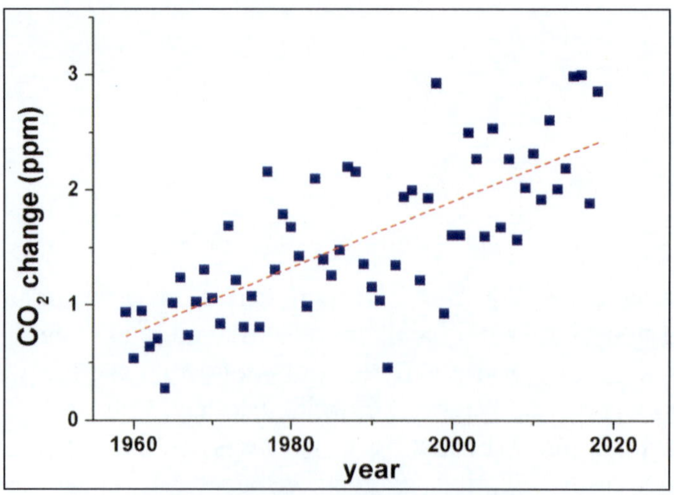

Recent year-to-year increase of atmospheric CO_2.

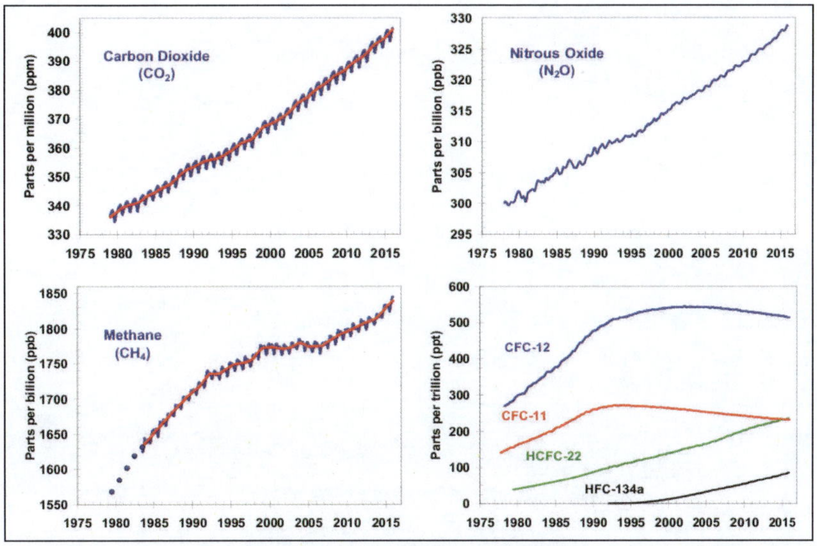

Major greenhouse gas trends.

Since the beginning of the Industrial Revolution, the concentrations of most of the greenhouse gases have increased. For example, the mole fraction of carbon dioxide has increased from 280 ppm to 415 ppm, or 120 ppm over modern pre-industrial levels. The first 30 ppm increase took place in about 200 years, from the start of the Industrial Revolution to 1958; however the next 90 ppm increase took place within 56 years, from 1958 to 2014.

Recent data also shows that the concentration is increasing at a higher rate. In the 1960s, the average annual increase was only 37% of what it was in 2000 through 2007.

Total cumulative emissions from 1870 to 2017 were 425±20 GtC (1539 $GtCO_2$) from fossil fuels and industry, and 180±60 GtC (660 $GtCO_2$) from land use change. Land-use change, such as deforestation, caused about 31% of cumulative emissions over 1870–2017, coal 32%, oil 25%, and gas 10%.

Today, the stock of carbon in the atmosphere increases by more than 3 million tonnes per annum (0.04%) compared with the existing stock. This increase is the result of human activities by burning fossil fuels, deforestation and forest degradation in tropical and boreal regions.

The other greenhouse gases produced from human activity show similar increases in both amount and rate of increase. Many observations are available online in a variety of Atmospheric Chemistry Observational Databases.

Anthropogenic Greenhouse Gases

Since about 1750 human activity has increased the concentration of carbon dioxide and other greenhouse gases. Measured atmospheric concentrations of carbon dioxide are currently 100 ppm higher than pre-industrial levels. Natural sources of carbon dioxide are more than 20 times greater than sources due to human activity, but over periods longer than a few years natural sources are closely balanced by natural sinks, mainly photosynthesis of carbon compounds by plants and marine plankton. As a result of this balance, the atmospheric mole fraction of carbon dioxide remained between 260 and 280 parts per million for the 10,000 years between the end of the last glacial maximum and the start of the industrial era.

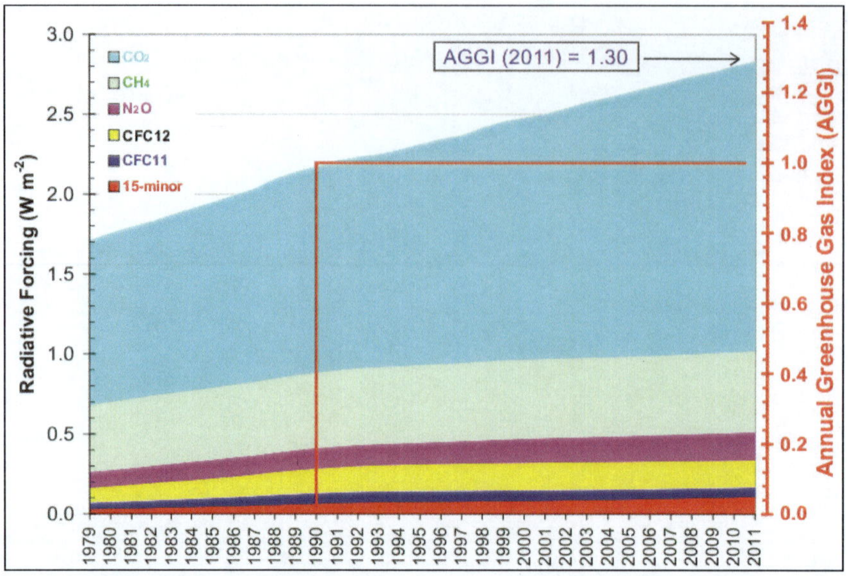

This graph shows changes in the annual greenhouse gas index (AGGI) between 1979 and 2011. The AGGI measures the levels of greenhouse gases in the atmosphere based on their ability to cause changes in Earth's climate.

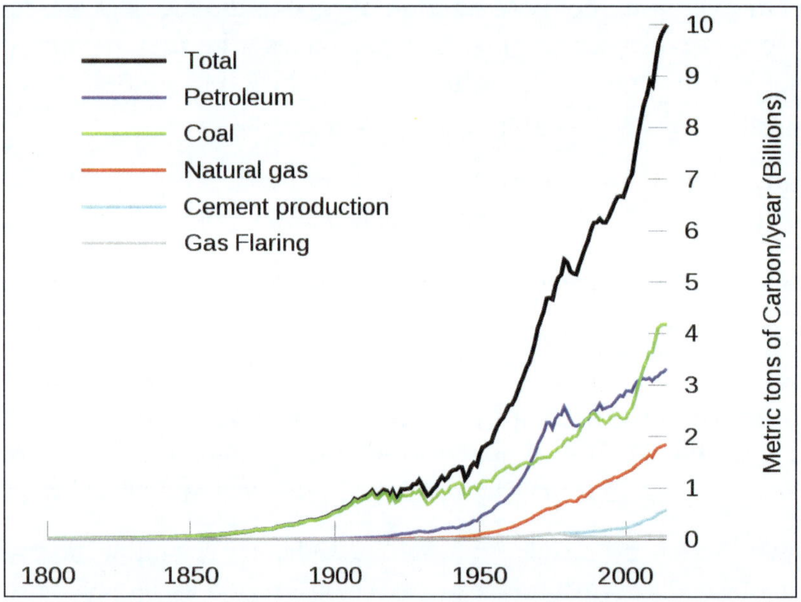

Modern global CO_2 emissions from the burning of fossil fuels.

It is likely that anthropogenic (i.e., human-induced) warming, such as that due to elevated greenhouse gas levels, has had a discernible influence on many physical and biological systems. Future warming is projected to have a range of impacts, including sea level rise, increased frequencies and severities of some extreme weather events, loss of biodiversity, and regional changes in agricultural productivity.

The main sources of greenhouse gases due to human activity are:

- Burning of fossil fuels and deforestation leading to higher carbon dioxide concentrations in the air. Land use change (mainly deforestation in the tropics) account for up to one third of total anthropogenic CO_2 emissions.

- Livestock enteric fermentation and manure management, paddy rice farming, land use and wetland changes, man-made lakes, pipeline losses, and covered vented landfill emissions leading to higher methane atmospheric concentrations. Many of the newer style fully vented septic systems that enhance and target the fermentation process also are sources of atmospheric methane.
- Use of chlorofluorocarbons (CFCs) in refrigeration systems, and use of CFCs and halons in fire suppression systems and manufacturing processes.
- Agricultural activities, including the use of fertilizers, that lead to higher nitrous oxide (N_2O) concentrations.

Mean greenhouse gas emissions for different food types	
Food Types	Greenhouse Gas Emissions (g CO_2-Ceq per g protein)
Ruminant Meat	62
Recirculating Aquaculture	30
Trawling Fishery	26
Non-recirculating Aquaculture	12
Pork	10
Poultry	10
Dairy	9.1
Non-trawling Fishery	8.6
Eggs	6.8
Starchy Roots	1.7
Wheat	1.2
Maize	1.2
Legumes	0.25

The seven sources of CO_2 from fossil fuel combustion are (with percentage contributions):

Seven main fossil fuel combustion sources	Contribution (%)
Liquid fuels (e.g., gasoline, fuel oil)	36%
Solid fuels (e.g., coal)	35%
Gaseous fuels (e.g., natural gas)	20%
Cement production	3 %
Flaring gas industrially and at wells	< 1%
Non-fuel hydrocarbons	< 1%
"International bunker fuels" of transport not included in national inventories	4 %

Carbon dioxide, methane, nitrous oxide (N_2O) and three groups of fluorinated gases (sulfur hexafluoride (SF_6), hydrofluorocarbons (HFCs), and perfluorocarbons (PFCs)) are the major anthropogenic greenhouse gases, and are regulated under the Kyoto Protocol international treaty, which came into force in 2005. Emissions limitations specified in the Kyoto Protocol expired in 2012. The

Cancún agreement, agreed on in 2010, includes voluntary pledges made by 76 countries to control emissions. At the time of the agreement, these 76 countries were collectively responsible for 85% of annual global emissions.

Although CFCs are greenhouse gases, they are regulated by the Montreal Protocol, which was motivated by CFCs' contribution to ozone depletion rather than by their contribution to global warming. Note that ozone depletion has only a minor role in greenhouse warming, though the two processes often are confused in the media. On 15 October 2016, negotiators from over 170 nations meeting at the summit of the United Nations Environment Programme reached a legally binding accord to phase out hydrofluorocarbons (HFCs) in an amendment to the Montreal Protocol.

Greenhouse Gases Emissions per Sector

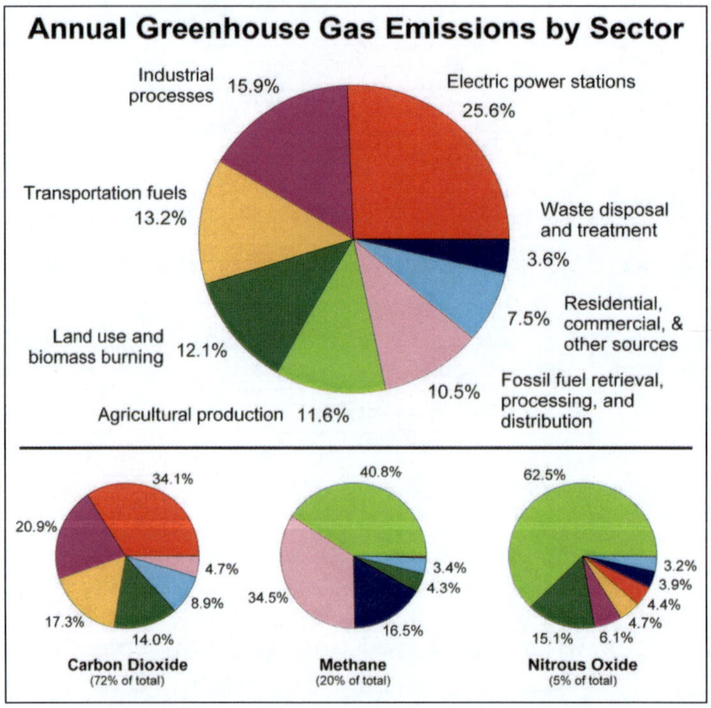

This figure shows the relative fraction of anthropogenic greenhouse gases coming from each of eight categories of sources, as estimated by the Emission Database for Global Atmospheric Research version 4.2, fast track 2010 project. These values are intended to provide a snapshot of global annual greenhouse gas emissions in the year 2010. The top panel shows the sum over all anthropogenic greenhouse gases, weighted by their global warming potential over the next 100 years. This consists of 72% carbon dioxide, 20% methane, 5% nitrous oxide and 3% other gases. Lower panels show the comparable information for each of these three primary greenhouse gases, with the same coloring of sectors as used in the top chart. Segments with less than 1% fraction are not labeled.

Electricity Generation

Electricity generation emits over a quarter of global greenhouse gases. Coal-fired power stations are the single largest emitter, with over 10 Gt CO_2 in 2018. Although much less polluting than coal plants, natural gas-fired power plants are also major emitters.

Tourism

According to UNEP, global tourism is closely linked to climate change. Tourism is a significant contributor to the increasing concentrations of greenhouse gases in the atmosphere. Tourism accounts for about 50% of traffic movements. Rapidly expanding air traffic contributes about 2.5% of the production of CO_2. The number of international travelers is expected to increase from 594 million in 1996 to 1.6 billion by 2020, adding greatly to the problem unless steps are taken to reduce emissions.

Trucking and Haulage

The trucking and haulage industry plays a part in production of CO_2, contributing around 20% of the UK's total carbon emissions a year, with only the energy industry having a larger impact at around 39%. Average carbon emissions within the haulage industry are falling—in the thirty-year period from 1977 to 2007, the carbon emissions associated with a 200-mile journey fell by 21 percent; NOx emissions are also down 87 percent, whereas journey times have fallen by around a third. Due to their size, HGVs often receive criticism regarding their CO_2 emissions; however, rapid development in engine technology and fuel management is having a largely positive effect.

Plastic

Plastic is produced mainly from Fossil fuels. Plastic manufacturing is estimated to use 8 percent of yearly global oil production. The EPA estimates as many as five ounces of carbon dioxide are emitted for each ounce of polyethylene terephthalate (PET) produced—the type of plastic most commonly used for beverage bottles, the transportation produce greenhouse gases also. Plastic waste emits carbon dioxide when it degrades. In 2018 research claimed that some of the most common plastics in the environment release the greenhouse gases methane and ethylene when exposed to sunlight in an amount that can affect the earth climate.

From the other side, if it is placed in a landfill, it becomes a carbon sink although biodegradable plastics have caused methane emissions. Due to the lightness of plastic versus glass or metal, plastic may reduce energy consumption. For example, packaging beverages in PET plastic rather than glass or metal is estimated to save 52% in transportation energy, if the glass or metal package is single use, of course.

In 2019 a new report "Plastic and Climate" was published. According to the report plastic will contribute greenhouse gases in the equivalent of 850 million tonnes of carbon dioxide (CO_2) to the atmosphere in 2019. In current trend, annual emissions will grow to 1.34 billion tonnes by 2030. By 2050 plastic could emit 56 billion tonnes of Greenhouse gas emissions, as much as 14 percent of the Earth's remaining carbon budget. The report says that only solutions which involve a reduction in consumption can solve the problem, while others like biodegradable plastic, ocean cleanup, using renewable energy in plastic industry can do little, and in some cases may even worsen it.

Pharmaceutical Industry

The pharmaceutical industry emitted 52 megatonnes of Carbon dioxide into the atmosphere in 2015. This is more than the automotive sector. However this analysis used the combined emissions of conglomerates which produce pharmaceuticals as well as other products.

Role of Water Vapor

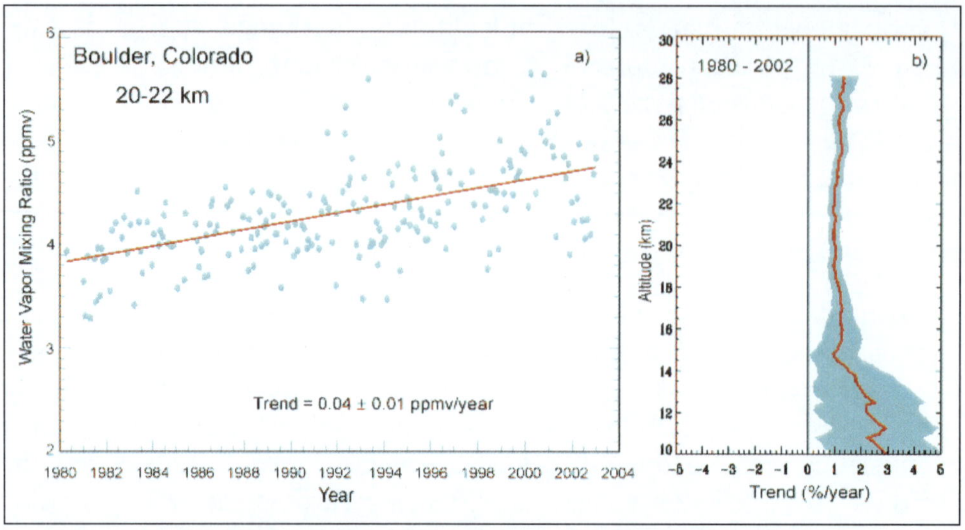

Increasing water vapor in the stratosphere at Boulder, Colorado.

Water vapor accounts for the largest percentage of the greenhouse effect, between 36% and 66% for clear sky conditions and between 66% and 85% when including clouds. Water vapor concentrations fluctuate regionally, but human activity does not directly affect water vapor concentrations except at local scales, such as near irrigated fields. Indirectly, human activity that increases global temperatures will increase water vapor concentrations, a process known as water vapor feedback. The atmospheric concentration of vapor is highly variable and depends largely on temperature, from less than 0.01% in extremely cold regions up to 3% by mass in saturated air at about 32 °C.

The average residence time of a water molecule in the atmosphere is only about nine days, compared to years or centuries for other greenhouse gases such as CH_4 and CO_2. Thus, water vapor responds to and amplifies effects of the other greenhouse gases. The Clausius–Clapeyron relation establishes that more water vapor will be present per unit volume at elevated temperatures. This and other basic principles indicate that warming associated with increased concentrations of the other greenhouse gases also will increase the concentration of water vapor (assuming that the relative humidity remains approximately constant; modeling and observational studies find that this is indeed so). Because water vapor is a greenhouse gas, this results in further warming and so is a "positive feedback" that amplifies the original warming. Eventually other earth processes offset these positive feedbacks, stabilizing the global temperature at a new equilibrium and preventing the loss of Earth's water through a Venus-like runaway greenhouse effect.

Direct Greenhouse Gas Emissions

Between the period 1970 to 2004, greenhouse gas emissions (measured in CO_2-equivalent) increased at an average rate of 1.6% per year, with CO_2 emissions from the use of fossil fuels growing at a rate of 1.9% per year. Total anthropogenic emissions at the end of 2009 were estimated at 49.5 gigatonnes CO_2-equivalent. These emissions include CO_2 from fossil fuel use and from land use, as well as emissions of methane, nitrous oxide and other greenhouse gases covered by the Kyoto Protocol.

At present, the primary source of CO_2 emissions is the burning of coal, natural gas, and petroleum for electricity and heat.

Regional and National Attribution of Emissions

According to the Environmental Protection Agency (EPA), GHG emissions in the United States can be traced from different sectors.

There are several ways of measuring greenhouse gas emissions,Some variables that have been reported include:

- Definition of measurement boundaries: Emissions can be attributed geographically, to the area where they were emitted (the territory principle) or by the activity principle to the territory produced the emissions. These two principles result in different totals when measuring, for example, electricity importation from one country to another, or emissions at an international airport.
- Time horizon of different gases: Contribution of a given greenhouse gas is reported as a CO_2 equivalent. The calculation to determine this takes into account how long that gas remains in the atmosphere. This is not always known accurately and calculations must be regularly updated to reflect new information.
- What sectors are included in the calculation (e.g., energy industries, industrial processes, agriculture etc.): There is often a conflict between transparency and availability of data.
- The measurement protocol itself: This may be via direct measurement or estimation. The four main methods are the emission factor-based method, mass balance method, predictive emissions monitoring systems, and continuous emissions monitoring systems. These methods differ in accuracy, cost, and usability.

These measures are sometimes used by countries to assert various policy/ethical positions on climate change. The use of different measures leads to a lack of comparability, which is problematic when monitoring progress towards targets. There are arguments for the adoption of a common measurement tool, or at least the development of communication between different tools.

Emissions may be measured over long time periods. This measurement type is called historical or cumulative emissions. Cumulative emissions give some indication of who is responsible for the build-up in the atmospheric concentration of greenhouse gases.

The national accounts balance would be positively related to carbon emissions. The national accounts balance shows the difference between exports and imports. For many richer nations, such as the United States, the accounts balance is negative because more goods are imported than they are exported. This is mostly due to the fact that it is cheaper to produce goods outside of developed countries, leading the economies of developed countries to become increasingly dependent on services and not goods. We believed that a positive accounts balance would means that more production was occurring in a country, so more factories working would increase carbon emission levels.

Emissions may also be measured across shorter time periods. Emissions changes may, for example, be measured against a base year of 1990. 1990 was used in the United Nations Framework

Convention on Climate Change (UNFCCC) as the base year for emissions, and is also used in the Kyoto Protocol. A country's emissions may also be reported as a proportion of global emissions for a particular year.

Another measurement is of per capita emissions. This divides a country's total annual emissions by its mid-year population. Per capita emissions may be based on historical or annual emissions.

While cities are sometimes considered to be disproportionate contributors to emissions, per-capita emissions tend to be lower for cities than the averages in their countries.

From Land-use Change

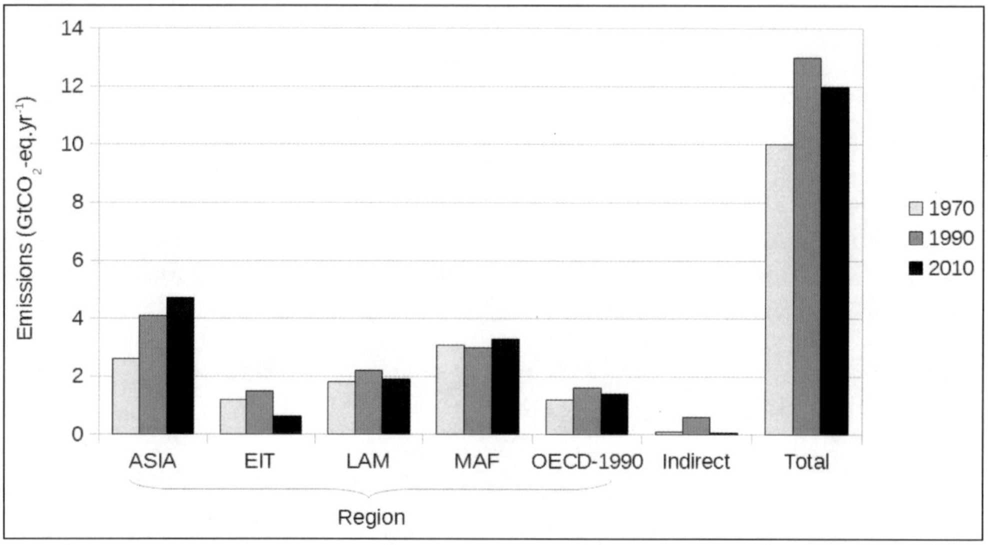

Greenhouse gas emissions from agriculture, forestry and other land use.

Land-use change, e.g., the clearing of forests for agricultural use, can affect the concentration of greenhouse gases in the atmosphere by altering how much carbon flows out of the atmosphere into carbon sinks. Accounting for land-use change can be understood as an attempt to measure "net" emissions, i.e., gross emissions from all sources minus the removal of emissions from the atmosphere by carbon sinks.

There are substantial uncertainties in the measurement of net carbon emissions. Additionally, there is controversy over how carbon sinks should be allocated between different regions and over time. For instance, concentrating on more recent changes in carbon sinks is likely to favour those regions that have deforested earlier, e.g., Europe.

Greenhouse Gas Intensity

Greenhouse gas intensity is a ratio between greenhouse gas emissions and another metric, e.g., gross domestic product (GDP) or energy use. The terms "carbon intensity" and "emissions intensity" are also sometimes used. Emission intensities may be calculated using market exchange rates (MER) or purchasing power parity (PPP). Calculations based on MER show large differences in intensities between developed and developing countries, whereas calculations based on PPP show smaller differences.

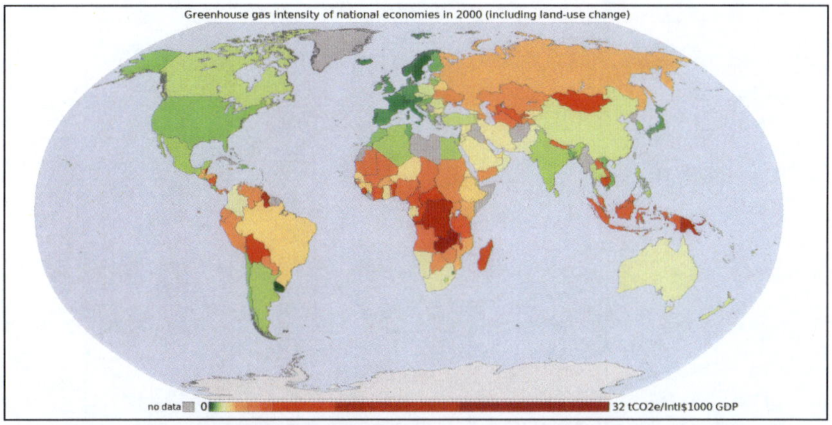

Greenhouse gas intensity in the year 2000, including land-use change.

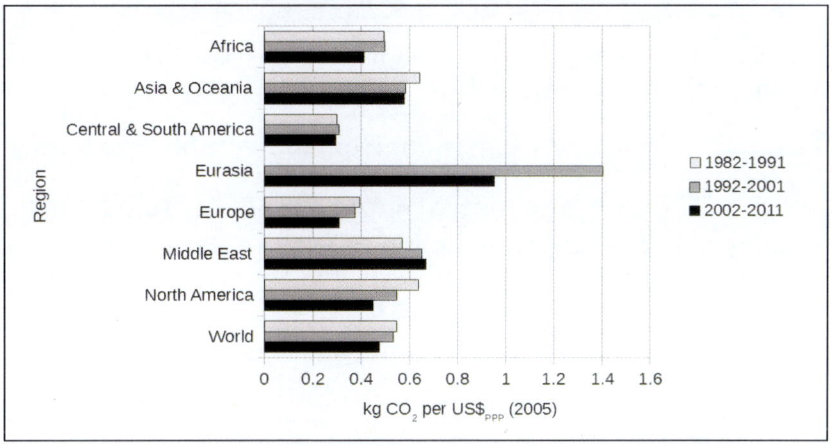

Carbon intensity of GDP (using PPP) for different regions.

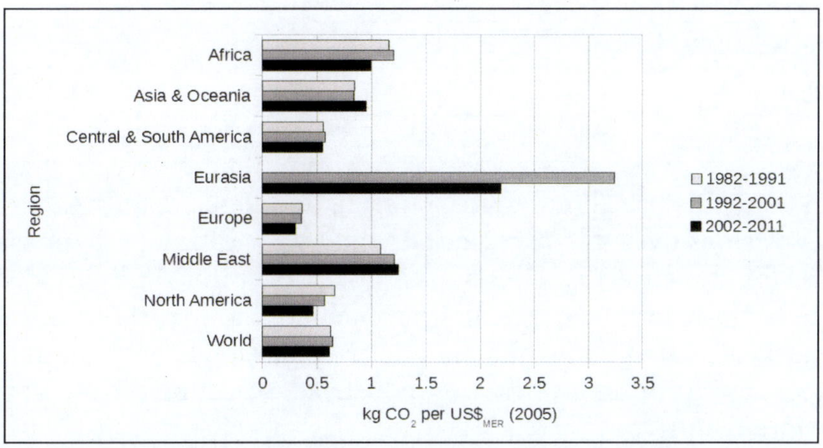

Carbon intensity of GDP (using MER) for different regions.

Cumulative and Historical Emissions

Cumulative energy-related CO_2 emissions between the years 1850–2005 grouped into low-income, middle-income, high-income, the EU-15, and the OECD countries.

Cumulative energy-related CO_2 emissions between the years 1850–2005 for individual countries.

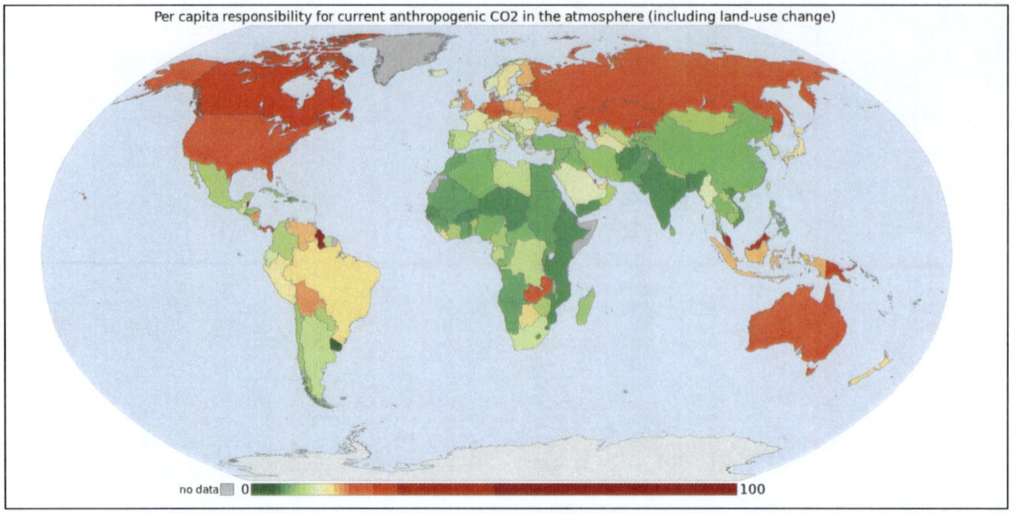

Map of cumulative per capita anthropogenic atmospheric CO_2 emissions by country. Cumulative emissions include land use change, and are measured between the years 1950 and 2000.

Regional trends in annual per capita CO_2 emissions from fuel combustion between 1971 and 2009.

Cumulative anthropogenic (i.e., human-emitted) emissions of CO_2 from fossil fuel use are a major cause of global warming, and give some indication of which countries have contributed most to human-induced climate change.

Top-5 historic CO_2 contributors by region over the years 1800 to 1988 (in %)		
Region	Industrial CO_2	Total CO_2
OECD North America	33.2	29.7
OECD Europe	26.1	16.6
Former USSR	14.1	12.5
China	5.5	6.0
Eastern Europe	5.5	4.8

The table above to the left is based on Banuri. Overall, developed countries accounted for 83.8% of industrial CO_2 emissions over this time period, and 67.8% of total CO_2 emissions. Developing countries accounted for industrial CO_2 emissions of 16.2% over this time period, and 32.2% of total CO_2 emissions. The estimate of total CO_2 emissions includes biotic carbon emissions, mainly from deforestation. Banuri calculated per capita cumulative emissions based on then-current population. The ratio in per capita emissions between industrialized countries and developing countries was estimated at more than 10 to 1.

Including biotic emissions brings about the same controversy regarding carbon sinks and land-use change. The actual calculation of net emissions is very complex, and is affected by how carbon sinks are allocated between regions and the dynamics of the climate system.

Non-OECD countries accounted for 42% of cumulative energy-related CO_2 emissions between 1890 and 2007. Over this time period, the US accounted for 28% of emissions; the EU, 23%; Russia, 11%; China, 9%; other OECD countries, 5%; Japan, 4%; India, 3%; and the rest of the world, 18%.

Changes since a Particular Base Year

Between 1970 and 2004, global growth in annual CO_2 emissions was driven by North America, Asia, and the Middle East. The sharp acceleration in CO_2 emissions since 2000 to more than a 3% increase per year (more than 2 ppm per year) from 1.1% per year during the 1990s is attributable to the lapse of formerly declining trends in carbon intensity of both developing and developed nations. China was responsible for most of global growth in emissions during this period. Localised plummeting emissions associated with the collapse of the Soviet Union have been followed by slow emissions growth in this region due to more efficient energy use, made necessary by the increasing proportion of it that is exported. In comparison, methane has not increased appreciably, and N_2O by 0.25% y^{-1}.

Using different base years for measuring emissions has an effect on estimates of national contributions to global warming. This can be calculated by dividing a country's highest contribution to global warming starting from a particular base year, by that country's minimum contribution to global warming starting from a particular base year. Choosing between base years of 1750, 1900, 1950, and 1990 has a significant effect for most countries. Within the G8 group of countries, it is most significant for the UK, France and Germany. These countries have a long history of CO_2 emissions.

Annual Emissions

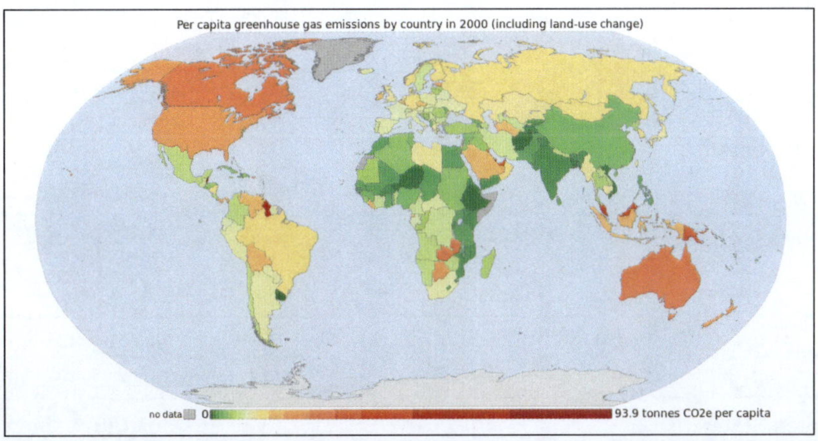

Per capita anthropogenic greenhouse gas emissions by country for the year 2000 including land-use change.

Annual per capita emissions in the industrialized countries are typically as much as ten times the average in developing countries. Due to China's fast economic development, its annual per capita emissions are quickly approaching the levels of those in the Annex I group of the Kyoto Protocol (i.e., the developed countries excluding the US). Other countries with fast growing emissions are South Korea, Iran, and Australia (which apart from the oil rich Persian Gulf states, now has the highest percapita emission rate in the world). On the other hand, annual per capita emissions of the EU-15 and the US are gradually decreasing over time. Emissions in Russia and Ukraine have decreased fastest since 1990 due to economic restructuring in these countries.

Energy statistics for fast growing economies are less accurate than those for the industrialized countries. For China's annual emissions in 2008, the Netherlands Environmental Assessment Agency estimated an uncertainty range of about 10%.

The greenhouse gas footprint refers to the emissions resulting from the creation of products or services. It is more comprehensive than the commonly used carbon footprint, which measures only carbon dioxide, one of many greenhouse gases.

2015 was the first year to see both total global economic growth and a reduction of carbon emissions.

Annual

In 2009, the annual top ten emitting countries accounted for about two-thirds of the world's annual energy-related CO_2 emissions.

Top-10 annual energy-related CO_2 emitters for the year 2009		
Country	% of global total annual emissions	Tonnes of GHG per capita
China	23.6	5.1
United States	17.9	16.9
India	5.5	1.4
Russia	5.3	10.8
Japan	3.8	8.6
Germany	2.6	9.2
Iran	1.8	7.3
Canada	1.8	15.4
South Korea	1.8	10.6
United Kingdom	1.6	7.5

Cumulative

Top-10 cumulative energy-related CO_2 emitters between 1850 and 2008		
Country	% of world total	Metric tonnes CO_2 per person
United States	28.5	1,132.7
China	9.36	85.4
Russia	7.95	677.2
Germany	6.78	998.9
United Kingdom	5.73	1,127.8
Japan	3.88	367
France	2.73	514.9
India	2.52	26.7
Canada	2.17	789.2
Ukraine	2.13	556.4

Embedded Emissions

One way of attributing greenhouse gas emissions is to measure the embedded emissions (also referred to as "embodied emissions") of goods that are being consumed. Emissions are usually measured according to production, rather than consumption. For example, in the main international treaty on climate change (the UNFCCC), countries report on emissions produced within their borders, e.g., the emissions produced from burning fossil fuels. Under a production-based accounting of emissions, embedded emissions on imported goods are attributed to the exporting, rather than the importing, country. Under a consumption-based accounting of emissions, embedded emissions on imported goods are attributed to the importing country, rather than the exporting, country.

Davis and Caldeira found that a substantial proportion of CO_2 emissions are traded internationally. The net effect of trade was to export emissions from China and other emerging markets to consumers in the US, Japan, and Western Europe. Based on annual emissions data from the year 2004, and on a per-capita consumption basis, the top-5 emitting countries were found to be (in CO_2 per person, per year): Luxembourg (34.7), the US (22.0), Singapore (20.2), Australia (16.7), and Canada (16.6). Carbon Trust research revealed that approximately 25% of all CO_2 emissions from human activities 'flow' (i.e., are imported or exported) from one country to another. Major developed economies were found to be typically net importers of embodied carbon emissions—with UK consumption emissions 34% higher than production emissions, and Germany (29%), Japan (19%) and the US (13%) also significant net importers of embodied emissions.

Relative CO_2 Emission from Various Fuels

One liter of gasoline, when used as a fuel, produces 2.32 kg (about 1300 liters or 1.3 cubic meters) of carbon dioxide, a greenhouse gas. One US gallon produces 19.4 lb (1,291.5 gallons or 172.65 cubic feet).

Mass of carbon dioxide emitted per quantity of energy for various fuels			
Fuel name	CO_2 emitted (lbs/10^6 Btu)	CO_2 emitted (g/MJ)	CO_2 emitted (g/kWh)
Natural gas	117	50.30	181.08
Liquefied petroleum gas	139	59.76	215.14
Propane	139	59.76	215.14
Aviation gasoline	153	65.78	236.81
Automobile gasoline	156	67.07	241.45
Kerosene	159	68.36	246.10
Fuel oil	161	69.22	249.19
Tires/tire derived fuel	189	81.26	292.54
Wood and wood waste	195	83.83	301.79
Coal (bituminous)	205	88.13	317.27
Coal (sub-bituminous)	213	91.57	329.65
Coal (lignite)	215	92.43	332.75
Petroleum coke	225	96.73	348.23
Tar-sand Bitumen			
Coal (anthracite)	227	97.59	351.32

Removal from the Atmosphere

Natural Processes

Greenhouse gases can be removed from the atmosphere by various processes, as a consequence of:

- A physical change (condensation and precipitation remove water vapor from the atmosphere).
- A chemical reaction within the atmosphere. For example, methane is oxidized by reaction with naturally occurring hydroxyl radical, OH• and degraded to CO_2 and water vapor (CO_2 from the oxidation of methane is not included in the methane Global warming potential). Other chemical reactions include solution and solid phase chemistry occurring in atmospheric aerosols.
- A physical exchange between the atmosphere and the other compartments of the planet. An example is the mixing of atmospheric gases into the oceans.
- A chemical change at the interface between the atmosphere and the other compartments of the planet. This is the case for CO_2, which is reduced by photosynthesis of plants, and which, after dissolving in the oceans, reacts to form carbonic acid and bicarbonate and carbonate ions.
- A photochemical change. Halocarbons are dissociated by UV light releasing Cl• and F• as free radicals in the stratosphere with harmful effects on ozone (halocarbons are generally too stable to disappear by chemical reaction in the atmosphere).

Negative Emissions

A number of technologies remove greenhouse gases emissions from the atmosphere. Most widely analysed are those that remove carbon dioxide from the atmosphere, either to geologic formations such as bio-energy with carbon capture and storage and carbon dioxide air capture, or to the soil as in the case with biochar. The IPCC has pointed out that many long-term climate scenario models require large-scale manmade negative emissions to avoid serious climate change.

Carbon Dioxide Equivalent

Carbon dioxide equivalent (CDE) and equivalent carbon dioxide (CO_2e and CO_2eq) are two related but distinct measures for describing how much global warming a given type and amount of greenhouse gas may cause, using the functionally equivalent amount or concentration of carbon dioxide (CO_2) as the reference.

Global Warming Potential

Carbon dioxide equivalency is a quantity that describes, for a given mixture and amount of greenhouse gas, the amount of CO_2 that would have the same global warming potential (GWP), when measured over a specified timescale (generally, 100 years). Carbon dioxide equivalency thus reflects the time-integrated radiative forcing of a quantity of *emissions* or rate of greenhouse gas emission—a *flow* into the atmosphere—rather than the instantaneous value of the radiative forcing of the *stock* (concentration) of greenhouse gases *in the atmosphere* described by CO_2e.

The carbon dioxide equivalency for a gas is obtained by multiplying the mass and the GWP of the gas. The following units are commonly used:

- By the UN climate change panel (IPCC): $n \times 10^9$ tonnes of CO_2 equivalent ($GtCO_2eq$).
- In industry: Million metric tonnes of carbon dioxide equivalents (MMTCDE).
- For vehicles: g of carbon dioxide equivalents/km (gCDE/km).

For example, the GWP for methane over 100 years is 34, and for nitrous oxide, 298. This means that emissions of 1 million tonnes of methane or nitrous oxide are equivalent to emissions of 34 or 298 million tonnes of carbon dioxide, respectively.

Equivalent Carbon Dioxide

Equivalent CO_2 (CO_2e) is the concentration of CO_2 that would cause the same level of radiative forcing as a given type and concentration of greenhouse gas. Examples of such greenhouse gases are methane, perfluorocarbons, and nitrous oxide. CO_2e is expressed as parts per million by volume, ppmv.

CO_2e calculation examples:

- The radiative forcing for pure CO_2 is approximated by $RF = \alpha \ln(C / C_0)$ where C is the present concentration, α is a constant, 5.35, and C_0 is the pre-industrial concentration, 280 ppm. Hence the value of CO_2e for an arbitrary gas mixture with a known radiative forcing is given by $C_0 \exp(RF / \alpha)$ in ppmv.
- To calculate the radiative forcing for a 1998 gas mixture, IPCC 2001 gives the radiative forcing (relative to 1750) of various gases as: CO_2=1.46 (corresponding to a concentration of 365 ppmv), CH_4=0.48, N_2O=0.15 and other minor gases =0.01 W/m2. The sum of these is 2.10 W/m2. Inserting this to the above formula, we obtain CO_2e = 412 ppmv.
- To calculate the CO_2e of the additional radiative forcing calculated from April 2016's updated data: $\sum$ RF(GHGs) = 3.3793, thus CO_2e = 280 $e^{3.3793/5.35}$ ppmv = 526.6 ppmv.

EFFECTS OF GLOBAL WARMING

Ocean Warming

Most of the additional heat generated by rising levels of CO_2 in the atmosphere is absorbed by the oceans. As a result, sea surface temperature is rising and already causing:

Thermal Expansion: As its temperature increases, seawater expands, contributing to sea level rise, changes in ocean circulation and higher seabed water temperature which may be damaging and pose the dangers described below:

Coastal Erosion: Rise in sea level, combined with other factors such as stronger wind events and loss of natural barriers protecting the coastline, result in increased coastal erosion endangering infrastructure, buildings and other facilities located in close proximity to the coastline.

Arctic Erosion: The Arctic ocean is warming, resulting in stronger storm activity and reduced sea ice formation, both contributing to erosion of coastlines hitherto kept stable by permafrost and sea ice reducing wave action. This causes increased exposure and thawing of methane (CH_4) bearing sediments and yedoma resulting in emission of this gas and its oxidation to CO_2, contributing to its rising presence in the atmosphere and further global warming.

Warmer bottom water: This is accelerating melting of ice enabling faster flow from glaciers discharging to the oceans and erosion of the marine ice sheet covering the West Antarctic archipelago – both causing sea level rise to accelerate and reducing stability of the ice sheet.

Warming seawater, particularly where shallow such as that covering the East Siberian Arctic Shelf, causes thawing of permafrost sediments containing CH_4, which is being released directly to the atmosphere, contributing to accelerated global warming.

Before and After. When corals are stressed by temperature they eject algae from their tissues, which give them their colour and die, losing the numerous varieties of fish which depend on them.

Coral Die-off: Reefs comprise a great variety of corals often growing in relatively shallow water. Coral reefs are weakened by human pollution making them susceptible to predation but are severely stressed or killed en masse by seawater temperature rising by 2C or more for 6-10 weeks. Their loss exposes adjacent coastlines, often low-lying, to erosion and flooding, destroys fish habitat and reduces fish catch for human consumption.

Ice Loss

Mountain glaciers store water which flows into rivers on which human populations depend for potable water, irrigation, food production, transport and generating energy, often in areas of dense population. These glaciers are storing less water and melting more rapidly so that in the future sufficient water may not be available for an expanding human population and its increasing demand for food and potable water.

Permafrost: Vast areas of land in the Arctic contain partly decomposed biota, sediments containing CH_4 produced from biota decomposition and yedoma. These lands are permanently frozen but global warming produces surface temperatures which result in it melting more rapidly and to greater depth. As it melts, it exposes biota which thaws, resuming decomposition and producing CH_4 much of which is converted to CO_2 through oxidation by methantrophic bacteria, then emitted to the atmosphere.

As permafrost melts the land subsides and becomes covered, in shallow water creating anoxic conditions in which methanotrophs are not active and this results in CH_4 from decaying biota and thawing yedoma being released to the atmosphere. CH_4 and CO_2 emitted from permafrost land thawing accelerate global warming. These emissions are already occurring and can not be safely controlled by human intervention.

Land Subsidence: Buildings and infrastructure built on permanently frozen land, particularly in Russia and Alaska, is put at risk when warming surface temperature causes permafrost to thaw and the land to subside. At risk are transport infrastructure, electricity supply, water and sewage mains, oil and gas pipelines and buildings and bridges – even entire cities. Land subsidence is likely to cause damage to the environment – eg. spills from ruptured pipelines. It may prove so costly to repair damage caused by subsidence as to force asset abandonment.

River flowing on surface of Greenland Ice Sheet into a moulin reaching the ice sheet base. Note discolouration of ice due to aerosol deposits of soot originating from forest and tundra fires, reducing albedo and causing faster ice melt.

Polar Ice Sheets: Global warming causes the surface of the Greenland Ice Sheet to melt more rapidly, resulting in rivers flowing on its surface, terminating in moulins through which they drain to bedrock. This intensely cold water lubricates the underside of the ice-sheet making it more mobile, before draining into the North Atlantic where it contributes to disruption of overturning circulation and flow of the Gulf Stream.

Ice Melt: Disruption of overturning circulation traps warmer water on the seabed causing the West Antarctic marine ice sheet to melt at its base, contributing to its instability. Warm seawater penetrates polar glaciers eroding ice blockages, enabling glaciers to discharge ice at faster rates resulting in ice sheets becoming less stable, as evidenced in Greenland and West Antarctica. This contributes to faster sea level rise which increases the risk of coastal erosion and flooding.

Sea Level Rise

We know that thermal expansion caused by ocean warming and loss of mass from ice sheets and glaciers are the primary causes of sea level rise. Less certain is the speed with which these causes take effect.

Many climate scientists specialising in this area, notably those contributing to IPCC Assessment Reports, are of the view that these are relatively slow processes indicating sea level rise of 0.52-0.98 metres by 2100. Others, including leading specialists in this field, point to evidence showing more rapid loss of ice, producing much faster, multi-metre sea level riseover the same period because of accelerating loss of mass from both the West Antarctic and Greenland ice caps, primarily as a result of increased glacier discharge rates.

Many of the worlds' cities are located on low coastal land which is very vulnerable to a multi-metre rise in sea level. Australia's Gold Coast with hundreds of kilometres of canals connected to the sea, Miami in Florida, even mega cities such as Shanghai would sustain heavy damage from sea level rise. Island nations such as Kiribati and the Maldives would be flooded, forcing abandonment. A rise of at least 2 metres is now thought possible by 2100.

World-wide the value of coastal property is likely plummet by hundreds of millions of dollars since there is no protection from rising seas or the increasingly severe storms which they will produce. Several hundred million people may have to retreat from coastal areas threatened by rising sea level this century.

Effects of Global Warming on Marine Mammals

The effect of global warming on marine mammals is a growing concern. Many of the effects of global warming are currently unknown due to unpredictability, but many are becoming increasingly evident today. Some effects are very direct such as loss of habitat, temperature stress, and exposure to severe weather. Other effects are more indirect, such as changes in host pathogen associations, changes in body condition because of predator–prey interaction, changed in exposure to toxins, and increased human interactions. Marine mammals that have been affected by climate change include walruses, seals, and polar bears.

Marine mammals have evolved to live in oceans, but climate change is affecting their natural habitat. Some species may not adapt fast enough, which might lead to their extinction.

During the last century, the global average land and sea surface temperature has increased due to an increased greenhouse effect from human activities. Warming has even reached depths of more than 700 meters (30% of ocean warming over the past decade has occurred in the deeper oceans below 700 meters). Many marine mammal species require specific temperature ranges to survive. Ocean warming will therefore lead to increased species migration, as endangered species look for a more suitable habitat. If a species cannot successfully migrate to a suitable environment, unless it learns to adapt to rising ocean temperatures, it will face extinction. Sea level rise is also important when assessing the impacts of global warming on marine mammals, since it affects coastal environments that marine mammals species rely.

Primary Productivity

Changes in temperatures change the location of areas with high primary productivity. Primary producers, such as plankton, are the main food source for marine mammals such as some whales. Species migration will therefore be directly affected by locations of high primary productivity. Water temperature changes also affect ocean turbulence, which has a major impact on the dispersion of plankton and other primary producers. Due to global warming and increased glacier melt, Thermohaline circulation patterns may be altered by increasing amounts of freshwater released into oceans and, therefore, changing ocean salinity. Thermohaline circulation is responsible for bringing up cold, nutrient-rich water from the depths of the ocean, a process known as upwelling.

Specific Marine Life and Mammals being Impacted

Polar bears are one of the marine mammals that are most at risk due to climate change. The biggest issue for polar bears related to climate change is the melting of ice as a result of increasing temperatures. When the ice melts, polar bears lose their habitat and food sources. Although polar bears have been known to eat more than 80 species of animals, most of their diet consists of seals, which are also endangered by global warming. There have been an increasing number of polar bear drownings because they become exhausted by having to swim farther to find ice or prey.

Not only are marine mammals impacted by climate change but so is other marine life. An example of this could be Coral. When coral is introduced to warming ocean temperatures changes, runoff and pollution, overexposure to sunlight extremely low tides and other stresses, the coral will expel an algae growing on them. They have a symbiotic relationship with the algae. When the coral expels the algae it becomes bleached or "completely white". This is called Coral Bleaching. The coral then become more vulnerable to disease and death. Coral reefs make up a large part of our oceans ecosystems that are teaming with life. One of the main homes of coral are in coral reefs. Climate change has caused a synergy between ocean acidification, ocean warming temperatures, and expansions of the oxygen minimum zone that will increase a sensitivity and vulnerability, that will move coral reef resources toward extinction.

Effects of Global Warming on Human Health

The effects of global warming include its effects on human health. The observed and projected increased frequency and severity of climate related impacts will further exacerbate the effects on human health. The various effects on individuals and populations are:

- Extreme heat: If high temperatures, especially when combined with high relative humidity, persist for several days (heat waves), and if nighttime temperatures do not drop, extreme heat can be a killer. Of all climate-related projections by scientists, rising temperatures are the most robust. Higher temperatures are also the most influenced by human behavior: the fewer heat-trapping emissions we release into the atmosphere, the cooler we can keep our planet. Because winter temperatures are rising faster than summer ones, cold-related deaths are likely to decline.
- "Natural" disasters: Projected changes in temperature and precipitation under global warming are likely to lead to other effects that threaten human health and safety. For example, changing precipitation patterns and prolonged heat can create drought, which can cause forest and peat fires, putting residents and firefighters in danger. However, a warming atmosphere also holds more moisture, so the chance of extreme rainfall and flooding continues to rise in some regions with rain or snow. In many heavily populated areas, sea-level rise is more likely to put people in the path of storm surges and coastal flooding. Warmer ocean waters may spawn more intense tropical hurricanes and typhoons while

ocean cycles continue to be a factor in the frequency of tropical cyclones.

- Poor air quality: Three key ingredients—sunlight, warm air, and pollution from power plants and cars burning coal and gasoline—combine to produce ground-level ozone (smog), which humans experience as poor air quality. Higher air temperatures increase smog, if sunlight, fossil fuel pollution, and air currents remain the same.

- Allergens and other nuisances: Warmer temperatures and higher concentrations of carbon dioxide in the atmosphere stimulate some plants to grow faster, mature earlier, or produce more potent allergens. Common allergens such as ragweed seem to respond particularly well to higher concentrations of CO_2, as do pesky plants such as poison ivy. Allergy-related diseases rank among the most common and chronic illnesses that can lead to lower productivity.

- Spreading diseases: Scientists expect a warmer world to bring changes in "disease vectors"—the mechanisms that spread some diseases. Insects previously stopped by cold winters are already moving to higher latitudes (toward the poles). Warmer oceans and other surface waters may also mean severe cholera outbreaks and harmful bacteria in certain types of seafood. Still, changes in land use and the ability of public health systems to respond make projecting the risk of vector-borne disease particularly difficult.

People do not bear the health risks of climate change equally because:

- Climate trends differ by region: People who live in floodplains, for example, are more likely to see river or coastal flooding. Similarly, people who live in regions with poor air quality today are at greater risk from poor air quality days in the future.

- Some people are more vulnerable to illness or death: Young children, the elderly, and those who are already ill are less able to withstand high temperatures and poor air quality, for example. Temperature extremes and smog hit people with heart and respiratory diseases, including asthma, particularly hard.

- Wealthy nations are more likely to adapt to projected climate change and recover from climate-related disasters than poor countries. Even within nations, less economically fortunate individuals are more vulnerable because they are less likely to have air conditioning and well-insulated homes, and because they have fewer resources to escape danger.

Better planning—through investments in infrastructure and public health strategies—can help communities become more resilient in a warming world. However, the costs of coping with health risks linked to severe climate change are often higher than the costs of curbing heat-trapping emissions in the first place.

References

- Mcneill, Leila. "This Lady Scientist Defined the Greenhouse Effect But Didn't Get the Credit, Because Sexism". Smithsonian. Retrieved 2019-04-16

- Global-warmin, science: britannica.com, Retrieved 4 June, 2019

- Held, Isaac M.; Soden, Brian J. (November 2000). "Water Vapor Feedback and Global Warming". Annual Review of Energy and the Environment. 25: 441–475. Citeseerx 10.1.1.22.9397. Doi:10.1146/annurev.energy.25.1.441

- Health, global-warming-effects: climatehotmap.org, Retrieved 5 July, 2019
- Charles J. Kibert (2016). "Background". Sustainable Construction: Green Building Design and Delivery. Wiley. ISBN 978-1119055327
- Global-Warming-Effects: skepticalscience.com, Retrieved 6 August, 2019
- Simmonds, Mark P.; Isaac, Stephen J. (5 March 2007). "The impacts of climate change on marine mammals: early signs of significant problems". Oryx. 41 (1): 19–26. Doi:10.1017/s0030605307001524

5

Climate Change Mitigation

Climate change mitigation is the process of reducing severity and negative effects of climate change. Some of its aspects include greenhouse gas removal, carbon dioxide removal, CCS and climate change mitigation, low-carbon diet, etc. The diverse aspects of climate change mitigation in the current scenario have been thoroughly discussed in this chapter.

The term mitigation refers to efforts to cut or prevent the emission of greenhouse gases - limiting the magnitude of future warming. It may also encompass attempts to remove greenhouse gases from the atmosphere.

It differs from climate change adaptation, which refers to the actions taken to manage the unavoidable impacts of climate change.

Mitigation may require us to use new technologies, clean energy sources, change people's behaviour, or make older technology more energy efficient.

Switching to low-carbon energy sources such as wind power, solar, geothermal, hydroelectric or nuclear represents one of the major strategies for lowering the emissions of greenhouse gases in the atmosphere.

The take-up of renewable energy - from sources that do not deplete over time - is growing. Figures from the UN show that renewable energy raised its share of total energy generation last year from 7.8% to 8.5%. But they also reveal that investment fell by 14%.

These technologies still face barriers related to capital costs (the upfront costs of building the plant and maintaining it), financing, public perception and a longstanding dependence of markets and institutions on fossil fuels.

But the IPCC's third report says many renewable energy technologies have made progress in both performance and cost and that their role in reducing air pollution and providing energy security outweighs possible disadvantages.

The scale of the challenge is daunting - fossil fuels make up about 80% of all energy. But technologies such as carbon capture and storage (CCS) could help reduce their impact. One form of CCS involves chemically capturing the carbon dioxide from a power station flue, and then piping it underground so that the invisible gas is contained in rock formations without leaking.

Existing oil fields, un-mineable coal seams and underground salty aquifers are all among the geological sites that are considered suitable for CO_2 storage. But while CCS could, in theory, limit the amount of carbon going into the atmosphere, it doesn't do much for the CO_2 already there.

Renewable energy such as solar power remains key, but faces barriers to take up.

Greening urban areas can also make a difference. Cities are home to half the planet's population, and are responsible for three-quarters of energy consumption and 80% of carbon emissions. Retro-fitting buildings to make them more energy efficient and cutting the impact of transport emissions represent some of the strategies for doing this.

Tackling waste is also an issue. About 11.2 billion metric tonnes of solid waste is currently being collected around the world every year, and the organic portion that decays is contributing around 5% to global greenhouse gas emissions. Experts would like to reduce its impact primarily by producing less but also recycling more and treating waste in a way that is less harmful to the environment or even using it as a sustainable energy fuel source.

Mitigation also extends to the protection of natural carbon "sinks" like the forests or oceans. New sinks can be created through, for example, forest regeneration. Geo-engineering is one controversial area that has gathered momentum in recent years. It requires the deliberate intervention in the climate system with the aim of curbing global warming.

One example is Solar Radiation Management (SRM), which involves reflecting more of the Sun's rays away from the planet back into space. This could be done by pumping sulphur aerosols into the high reaches of the atmosphere, where they would have similar reflective properties to the ash released naturally by volcanoes.

The IPCC's third report also considers a little-known technique called Bio-Energy Carbon Capture and Storage (Beccs). As the name suggests, this would combine the burning of biomass, such as wood, for energy and then piping the CO_2 into rocks. In theory, you could begin to remove emissions that have already been accumulated in the atmosphere.

But critics of this approach warn that the technology is unproven and could give reluctant countries an excuse not to cut emissions.

GREENHOUSE GAS REMOVAL

Greenhouse gas removal projects are a type of climate engineering that seek to remove greenhouse gases from the atmosphere, and thus they tackle the root cause of global warming. These techniques either directly remove greenhouse gases, or alternatively seek to influence natural processes to remove greenhouse gases indirectly. The discipline overlaps with carbon capture and storage and carbon sequestration, and some projects listed may not be considered to be climate engineering by all commentators, instead being described as mitigation.

Carbon Sequestration

A wide range of techniques for carbon sequestration exist. These range from ideas to remove CO_2 from the atmosphere (carbon dioxide air capture), flue gases (carbon capture and storage) and by preventing carbon in biomass from re-entering the atmosphere, such as with Bio-energy with carbon capture and storage (BECCS).

Pyrogenic Carbon Capture and Storage

Pyrogenic carbon capture and storage is discussed as a promising technology of greenhouse gas removal. Pyrolysis is described as "the thermal treatment of biomass at 350 °C–900 °C in an oxygen-deficient atmosphere. Three main carbonaceous products are generated during this process, which can be stored subsequently in different ways to produce (negative emissions): A solid biochar as soil amendment, a pyrolytic liquid (bio-oil) pumped into depleted fossil oil repositories, and permanent-pyrogas (dominated by the combustible gases CO, H_2 and CH_4) that may be transferred as CO_2 to geological storages after combustion".

A different procedure has been proposed by Esrafilzadeh that makes it possible to create bio char under room temperature, using liquid metal electrocatalysts.

Chlorofluorocarbon Photochemistry

Atmospheric chlorofluorocarbon (CFC) removal is an idea which suggests using lasers to break up CFCs, an important family of greenhouse gases, in the atmosphere.

Methane Removal

Methane potentially poses major challenges for remediation. It is around 20 times as powerful a greenhouse gas as CO_2. Large quantities may be outgassed from permafrost and clathrates as a result of global warming, notably in the Arctic.

There are existing climate engineering proposals. Methane is removed by several natural processes, which can be enhanced.

- Chemical decomposition: Reaction with hydroxyl radicals produced from photochemical decomposition of ozone in the stratosphere.
- Biological decomposition: By methanotrophs in soils and water.

CARBON DIOXIDE REMOVAL

Carbon dioxide removal (CDR) refers to a group of technologies the objective of which is the large-scale removal of carbon dioxide from the atmosphere. Among such technologies are bio-energy with carbon capture and storage, biochar, ocean fertilization, enhanced weathering, and direct air capture when combined with storage. CDR is a different approach from removing CO_2 from the stack emissions of large fossil fuel point sources, such as power stations. The latter reduces emission to the atmosphere but cannot reduce the amount of carbon dioxide already in the atmosphere. As CDR removes carbon dioxide from the atmosphere, it 'creates' negative emissions that offset the emissions from small and dispersed point sources such as domestic heating systems, airplanes and vehicle exhausts. It is regarded by some as a form of climate engineering, while other commentators describe it as a form of carbon capture and storage or extreme mitigation. Whether CDR would satisfy common definitions of "climate engineering" or "geoengineering" usually depends upon the scale at which it would be undertaken.

The likely need for CDR has been publicly expressed by a range of individuals and organizations involved with climate change issues, including IPCC chief Rajendra Pachauri, the UNFCCC executive secretary Christiana Figueres, and the World Watch Institute. Institutions with major programs focusing on CDR include the Lenfest Center for Sustainable Energy at the Earth Institute, Columbia University, and the Climate Decision Making Center, an international collaboration operated out of Carnegie-Mellon University's Department of Engineering and Public Policy.

The mitigation effectiveness of air capture is limited by societal investment, land use, availability of geologic reservoirs, and leakage. The reservoirs are estimated to be sufficient for storing at least 545 gigatonnes of carbon (GtC). Storing 771 GtC would cause a 186 ppm atmospheric reduction. In order to return the atmospheric CO_2 content to 350 ppm we would need atmospheric reductions of 50 ppm each year and also to reduce current emissions by the equivalent of 2 ppm per year.

Carbon dioxide removal is different from reducing emissions, as the former produces an outlet of carbon dioxide from Earth's atmosphere, whereas the latter decreases the inlet of carbon dioxide to the atmosphere. Both have the same net effect, but for achieving carbon dioxide concentration levels below present levels, carbon dioxide removal is critical. Also for meeting higher concentration levels, carbon dioxide removal is increasingly considered to be crucial as it provides the only possibility to fill the gap between needed reductions to meet mitigation targets and global emission trends.

In the OECD Environmental Outlook to 2050 released at the 2011 United Nations Climate Change Conference, the authors commented on the need for negative emissions, stating "Achieving lower concentration targets (450 ppm) depends significantly on the use of BECCS".

A carbon dioxide sink such as a concentrated group of plants or any other primary producer that binds carbon dioxide into biomass, such as within forests and kelp beds, is not carbon negative, as sinks are not permanent. A carbon dioxide sink of this type moves carbon, in the form of carbon dioxide, from the atmosphere or hydrosphere to the biosphere. This process could be undone, for example by wildfires or logging.

Carbon dioxide sinks that store carbon dioxide in the Earth's crust by injecting it into the subsurface, or in the form of insoluble carbonate salts (mineral sequestration), are considered carbon negative. This is because they are removing carbon from the atmosphere and sequestering it indefinitely and presumably for a considerable duration (thousands to millions of years). However, Carbon Capture technology remains, at best, theoretical and is yet to reach more than 33% efficiency. Furthermore, this process could be rapidly undone, for example by earthquakes or mining.

Methods

Bio-energy with Carbon Capture and Storage

Bio-energy with carbon capture and storage, or BECCS, uses biomass to extract carbon dioxide from the atmosphere, and carbon capture and storage technologies to concentrate and permanently store it in deep geological formations.

BECCS is currently the only CDR technology deployed at full industrial scale, with 550 000 tonnes CO_2/year in total capacity operating, divided between three different facilities.

The Imperial College London, the UK Met Office Hadley Centre for Climate Prediction and Research, the Tyndall Centre for Climate Change Research, the Walker Institute for Climate System Research, and the Grantham Institute for Climate Change issued a joint report on carbon dioxide removal technologies as part of the *AVOID: Avoiding dangerous climate change* research program, stating that “Overall, of the technologies studied in this report, BECCS has the greatest maturity and there are no major practical barriers to its introduction into today’s energy system. The presence of a primary product will support early deployment”.

According to the OECD, “achieving lower concentration targets (450 ppm) depends significantly on the use of BECCS”.

Biochar

Biochar is created by the pyrolysis of biomass, and is under investigation as a method of carbon sequestration. Biochar is a charcoal that is used for agricultural purposes which also aids in carbon sequestration, the capture or hold of carbon. It is created using a process called pyrolysis, which is basically the act of high temperature heating biomass in an environment with low oxygen levels. What remains is a material known as char, similar to charcoal but is made through a sustainable process, thus the use of biomass. Biomass is organic matter produced by living organisms or recently living organisms, most commonly plants or plant based material. The offset of greenhouse gas (GHG) emission, if biochar were to be implemented, would be a maximum of 12%. This equates to about 106 metric tons of CO_2 equivalents. On a medium conservative level, it would be 23% less than that, at 82 metric tons. A study done by the UK Biochar Research Center has stated that, on a conservative level, biochar can store 1 gigaton of carbon per year. With greater effort in marketing and acceptance of biochar, the benefit could be the storage of 5–9 gigatons per year of carbon in biochar soils.

Enhanced Weathering

Enhanced weathering is a chemical approach to remove carbon dioxide involving land- or ocean-based techniques. One example of a land-based enhanced weathering technique is in-situ carbonation of silicates. Ultramafic rock, for example, has the potential to store from hundreds to thousands of years’ worth of CO_2 emissions, according to estimates. Ocean-based techniques involve alkalinity enhancement, such as grinding, dispersing, and dissolving olivine, limestone, silicates, or calcium hydroxide to address ocean acidification and CO_2 sequestration. Enhanced weathering is considered one of the least expensive geoengineering options. One example of a research project on the feasibility of enhanced weathering is the CarbFix project in Iceland.

Direct Air Capture (DAC)

Carbon dioxide can be removed from ambient air through chemical processes, sequestered, and stored. Traditional modes of carbon capture such as precombustion and postcombustion CO_2 capture from large point sources can help slow the rate of increase of the atmospheric CO_2 concentration, but only the direct removal of CO_2 from the air, or direct air capture (DAC), can actually reduce the global atmospheric CO_2 concentration if combined with long-term storage of CO_2.

DAC relying on amine-based absorption demands significant water input. It was estimated, that to capture 3.3 Gigatonnes of CO_2 a year would require 300 km^3 of water, or 4% of the water used for irrigation. On the other hand, using sodium hydroxide needs far less water, but the substance itself is highly caustic and dangerous.

DAC also requires much greater energy input in comparison to traditional capture from point sources, like flue gas, due to the low concentration of CO_2. The theoretical minimum energy required to extract CO_2 from ambient air is about 250 kWh per tonne of CO_2, while capture from natural gas and coal power plants requires respectively about 100 and 65 kWh per tonne of CO_2.

Ocean Fertilization

Ocean fertilization or ocean nourishment is a type of climate engineering based on the purposeful introduction of nutrients to the upper ocean to increase marine food production and to remove carbon dioxide from the atmosphere. A number of techniques, including fertilization by iron, urea and phosphorus have been proposed.

Wooden Building Construction

In their "roadmap for rapid decarbonization", Rockström proposed in the Science that after 2030, all building construction must be carbon-neutral or carbon-negative. The construction industry must either use emissions-free concrete and steel or replace those materials with zero- or negative-emissions substances such as wood, stone, and carbon fiber but not tools of daily use. Especially wooden building constructions are a promising way to store emissions, as wood mainly contains carbon, grows faster due to the higher CO_2 concentration and is established as a building material for several centuries. Projects like W350 in Tokyo and Mjøstårnet in Brumunddal demonstrate that even skyscrapers could potentially be built from wood and, thus, store large amounts of carbon.

Example CO_2 Scrubbing Chemistry

Calcium Oxide

Calcium oxide (quicklime) will absorb CO_2 from atmospheric air mixed with steam at 400 °C (forming calcium carbonate) and release it at 1,000 °C. This process, proposed by A. Steinfeld, can be performed using renewable energy from thermal concentrated solar power. Quicklime is made by heating limestone to release the CO_2 within it. Quicklime is mixed with sand for brick building as mortar, where it hardens by absorption of CO_2.

Sodium Hydroxide

Zeman and Lackner outlined a specific method of air capture using sodium hydroxide. Carbon Engineering, a Calgary, Alberta firm founded in 2009 and partially funded by Bill Gates, is developing a process to capture carbon dioxide using a solution of potassium hydroxide mixed with some water at their pilot plant. They hope to create and sell synthetic fuels at a cost of $100 a ton. They have partnered with Greyrock.

Direct Air Capture (DAC) Process Example using NaOH

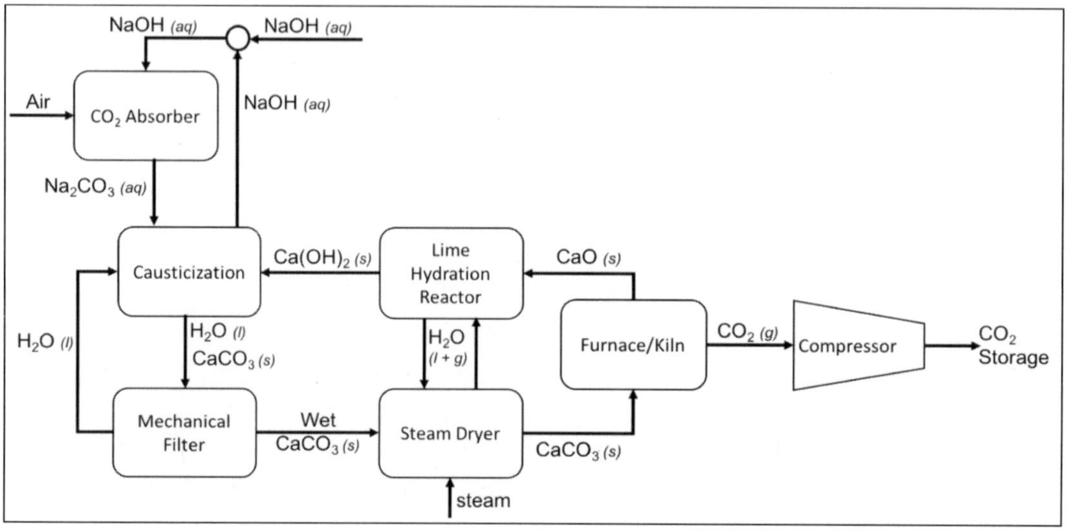

Main steps in a DCA process that uses NaOH (caustic soda) as the absorber.

Among the technologies studied for direct air capture (DAC), the use of aqueous hydroxide sorbents is one of the most promising approaches. In this process, CO_2 from the air is chemically dissolved into NaOH(aq) solution as Na_2CO_3; the Na_2CO_3 is then reacted with solid $Ca(OH)_2$, which regenerates the solvent and produces $CaCO_3$ crystals; lastly, heat is applied to the $CaCO_3$ crystals to produce pure CO_2 gas.

Air is pumped through the CO_2 absorber as the first step of this process. CO_2 absorber for DAC are designed either as a counter-current spray tower or as a counter-current thin-falling-film contractor to maximize the contact area between the air and the solvent and thus maximize the absorption driving force. The solvent is regenerated in the causticization unit by reacting the Na_2CO_3 with $Ca(OH)_2$, which also transfers the captured CO_2 to the form of $CaCO_3$ solid crystals. A mechanical filter is then used to separate the $CaCO_3$ crystals from the water. Since the crystals come out wet from the filter, they are dried in a steam dryer. Then the dry crystals are heated in a furnace to produce CaO and pure CO_2 gas. The CaO is then hydrated to regenerate the $Ca(OH)_2$ used for the causticization reaction. The pure CO_2 stream is then compressed and ready to be transported for geologic sequestration, EOR, or other commercial applications.

1 M NaOH (aq) is a typical solvent concentration because this concentration is limited by the causticization reaction that regenerates the solvent and it is not too far from the practical maximum of 2 M NaOH. The furnace/kiln can be powered renewably or by burning fuel on-site with pure oxygen produces in an on-site air separation unit.

NaOH is economically competitive with other absorbents--e.g., amines--used for DAC processes. DAC processes are energy intensive. Calcination (at the furnace) is the most energy intensive step of this process.

Risks, Problems and Criticisms

CDR is slow to act, and requires a long-term political and engineering program to effect. CDR is even slower to take effect on acidified oceans. In a Business as usual concentration pathway, the deep ocean will remain acidified for centuries, and as a consequence many marine species are in danger of extinction.

The Special 1.5°C IPCC report was very clear about CDR : "CDR deployed at scale is unproven and reliance on such technology is a major risk in the ability to limit warming to 1.5°C".

CCS AND CLIMATE CHANGE MITIGATION

Carbon capture and storage (CCS) is a method used to reduce the amount of anthropogenic emissions of carbon dioxide (CO_2) in the atmosphere in an attempt to limit the effects of global climate change. CCS is can be employed to achieve a number of goals regarding climate change mitigation, such as preventing average global temperature from reaching certain levels above the pre-industrial average. In December 2015, the Paris Agreement articulated a census to not exceed pre-industrial global temperatures by more than 2 °C and recognized that different countries would have different contributions to help realize this goal. Under the Paris Agreement, different scenarios and climate models were analyzed for different temperature goals considering a wide range of mitigation methods from a temperature goal of less than 2 °C to an upper limit of exactly 2 °C increase above the pre-industrial average.

The terms CCS and CCUS (Carbon Capture, Utilization, and Storage) are often used interchangeably. The difference between the two is the specified 'utilization' of the captured carbon and refers to its use for other applications, such as enhanced oil recovery (EOR), potentially making liquid fuel, or the manufacturing of useful consumer goods, such as plastics. Since both approaches capture emitted CO_2 and effectively store it, whether that be under-ground in geological formations or long-term trapping in material products, the two terms are often treated the same.

CCS is considered as a basis of one climate stabilization wedge, which is a proposed climate mitigation action to reduce approximately 1 billion tonnes of carbon emissions over 50 years.

CCS and Different Climate Models

Large scale CCS plays a crucial role in reaching climate change stabilization. According to the IPCC, the carbon emission patterns can greatly vary based on the uncertainty of human power consumption. A file regarding the fluctuations of greenhouse gas emissions is shown to the right. However, CCS' primarily role is to delay the shift from fossil fuels and thereby reducing transition costs. The implementation of default technology assumptions would cost 29-297% more over the century than efforts without CCS for a 430-480 ppm CO_2/yr scenario. The Paris agreement upholds a goal to reach no more than a 2.0 °C increase above pre-industrial temperatures. If the 2.0 °C goal is to be reached in time, CCS must be utilized to achieve net zero emissions by 2060-2070. After 2060-2070, negative emissions will need to be achieved to remain below the 2.0 °C target. The variations in methods depend heavily on the climate change model being used and the anticipated energy consumption patterns. It is widely agreed upon, however, that CCS would need to be utilized if there is to be any negative climate change mitigation.

CCS and 2.0 °C Target

The concept of a 2.0 °C came to light in the European Union of 1996 where the goal was to reduce the global temperature range relative pre-industrial levels. The decision of the 2 °C range was decided mostly on the evidence that many ecosystems are at risk if average global temperatures exceeded this limit. In order to limit the anthropogenic emissions such that there is no more than a 2 °C change relative to the periods between 1861-1880, carbon emissions would need to be limited to about 1000 GtC by 2100 since that period. However, by the end of 2011 about of half of the budget was already released (445 GtC) indicating that a lower budget is necessary.

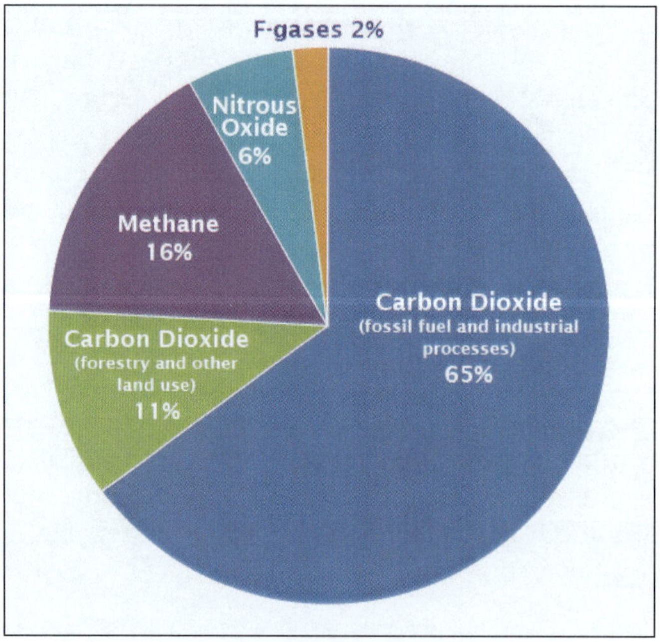

Global greenhouse gas emissions by gas type.

A distinctive path that aims for a 2.0 °C limit might have complications. The first complication involves the lack of positive feedback loops in IPCC climate models. These loops include reduction of ice sheet size, which would mean less sunlight is reflected and more is absorbed by the darker colored ground or water, and the potential release of greenhouse gases by thawing tundra. Since the lifetime of CO_2 in the climate atmosphere is so long,these feedback loops have to be taken into consideration. Another important factor to consider is that a 2.0 °C scenario necessitates tapping into alternative fossil fuels sources that are harder to obtain. Some examples of these methods are the exploitation of tar sands, tar shales, hydrofracking for oil and gas, coal mining, drilling in the

Arctic, Amazon, and deep ocean. Therefore, 2.0 °C scenarios result in more CO_2 produced per unit of usable energy. Further, the danger of extra released CH_4 via mining processes must be taken into account.

Different models are based on when the peak of carbon emissions happen on a global scale. Regarding the 2.0 °C scenario with respect to pre-industrial levels, possible approaches are short term and long term emission resolutions as well as the considering the cost effectiveness of different solutions to reduce carbon emissions. Short term goals are set to quantify progress towards the temperature goal. In a short term goal, looking ahead to the year 2020, the allowable carbon emissions must be between 41 and 55 $GtCO_2$ per year. The short term 2 °C scenario is not feasible without CCS.

Currently, greenhouse gas emissions would need to be reduced by 7 Gt of carbon equivalent each year by 2050 to achieve 2 °C stabilization. This requires power generation with CCS at 800 coal-fired power plants of 1 GW energy generation capacity,180 coal-synfuel plants, or natural gas plants worth 1,600 GW. In this scenario, one of the wedges, or 1 Gt of carbon is accounted for by CCUS. The cost of capturing CO_2 is estimated to be $500/tC. If the goal with the 2.0 °C is to store a total of 7 Gt carbon per year, then the collective amount needed to achieve this is around 3.5 trillion U.S. dollars per year. The economic demand needed to achieve this goal is high. This amount of money is equivalent to gross national incomes for countries such as Russia or United Kingdom, and it represents 18% of United States' 2017 gross national income (19.61 trillion dollars).

CCS and below 2.0 °C Target

Achieving below 2.0 °C Target

A change of temperature below 2 °C is, to certain extent, almost impossible to achieve due to the current carbon emission practices. The IPCC notes that it is difficult to assess a climate mitigation scenario that would limit average global temperature increase to only 1.5°C above pre-industrial levels. This is mainly due to the fact that few reliable multi-model studies have been conducted to thoroughly explore this scenario. Nevertheless, what few studies that have been done agree that mitigation technologies must be implemented immediately and scaled up quickly and reflect energy demand decrease. A change below 1 °C with respect to pre-industrial era is now inconceivable because by 2017 there was already an increase of 1 °C.

Because of the immediate inability to control the temperature at the 1 °C target, the next realistic target is 1.5 °C. There is enough confidence that past emissions alone (pre-industrial time) will not be enough to go beyond the 1.5 °C target. In other words, if all anthropogenic emissions were stopped today (reduced to zero), any increase beyond the 1 °C change for more than half of a degree before 2100 is unlikely. If anthropogenic emission are considered, the probability for the planet increasing for more than 1.5 °C before 2100 are high.Then, scenarios where the degree change is maintain below 1.5 °C are very challenging to achieve but not impossible.

For a below 2.0 °C target, Shared socioeconomic pathways (SSPs) had been developed adding a socio-economic dimension to the integrative work started by RCPs models.The advantage of using SSPs is that they incorporate social standards, fossil fuel use, geographical development, and high energy demand. SSPs also incorporate the use of six other models such as GCAM4,IMAGE,-MESSAGE-GLOBIOM,REMIND-MAgPIE. The combination of models and scenarios concluded that by 2050, annual CO_2 emissions are in the range between 9 and 13 billion tons of CO_2. All of the scenarios estimated that temperature will remain below 2.0 °C change with a 66% probability of success. To do so, a 1.9 W/m^2 within the year 2100 is necessary. Net zero GHG emissions have to be achieved between 2055-2075, and CO_2 emissions have to be in a range between 175 and 475 $GtCO_2$ between the years 2016-2100. All SSPs scenarios show a shift away from unabated fossil fuels, that is process without CCS.

Assumptions for below 2.0 °C Target

To achieve a 1.5 °C target before 2100, the following assumptions have to be considered; emissions have to peak by 2020 and decline after that,it will be necessary to reduce net CO_2 emissions to zero and negative emissions have to be a reality by the second half of the 21st century. For this assumptions to take place, CCS has to be implemented in factories that accompany the use of fossil fuels. Because emissions reduction has to be implemented more rigorously for a 1.5 °C target, methods such as BEECS, and natural climate solutions such as afforestation can be used to aim in the reduction of global emissions. BECCS is necessary to achieve a 1.5 °C. It is estimated by the models that with the help of BECCS, between 150-12000 $GtCO_2$ still have to be removed from the atmosphere.

Another negative emission strategy which includes CCS can also be approached through DACCS. Direct Air Carbon Capture and Sequestration (DACCS) is a carbon negative technology that utilizes solid amine based capture and it has proven to capture carbon dioxide from the air even though content of the air is much lower than of a flue gas from a coal plant. However, it would require renewable energies to power since approximately 400kJ of work is needed per mole of CO_2 capture. Furthermore, it is estimated that the total system cost is $1,000 per tonne of CO_2, according to an economic and energetic analysis from 2011.

Going forward in the utilization of models such as SSPss and RCP, feasibility of the model has to be to take into consideration. Feasibility includes concerns in various fields, such as geophysics, technology, economics, social acceptance, and politics, all of which can serve to facilitate or obstruct the carbon capture and sequestration of emissions needed in order to achieve the global temperature targets.Uncertainty in feasibility is especially a problem with more strict temperatures limits such as 1.5 °C. Real world feasibility of SSPs models, or any other models, in general are coarse approximations of reality.

LOW-CARBON DIET

A low-carbon diet refers to making lifestyle choices to reduce the greenhouse gas emissions (GHGe) resulting from consumption decisions. It is estimated that the U.S. food system is responsible for at least 20 percent of U.S. greenhouse gases. This estimate may be low, as it counts only direct sources of GHGe. Indirect sources, such as demand for products from other countries, are often not counted. A low-carbon diet minimizes the emissions released from the production, packaging, processing, transport, preparation and waste of food. Major tenets of a low-carbon diet include eating less industrial meat and dairy, eating less industrially produced food in general, eating food grown locally and seasonally, eating less processed and packaged foods and reducing waste from food by proper portion size, recycling or composting.

A 2014 study into the real-life diets of British people estimated their greenhouse gas footprints in terms of kilograms of carbon dioxide equivalent per day:

- 7.19 for high meat-eaters (≥100 g/day).
- 5.63 for medium meat-eaters (50–99 g/day).
- 4.67 for low meat-eaters (<50 g/day).
- 3.91 for fish-eaters.
- 3.81 for vegetarians.
- 2.89 for vegans.

In the U.S., the food system emits four of the greenhouse gases associated with climate change: carbon dioxide, methane, nitrous oxide, and chlorofluorocarbons. The burning of fossil fuels (such as oil and gasoline) to power vehicles that transport food for long distances by air, ship, truck and rail releases carbon dioxide (CO_2), the primary gas responsible for global warming. Chlorofluorocarbons (CFCs) are emitted from mechanical refrigerating and freezing mechanisms – both staples in food shipment and storage. Anthropogenic methane emission sources include agriculture (ruminants, manure management, wetland rice production), various other industries and landfills. Anthropogenic nitrous oxide sources include fertilizer, manure, crop residues and nitrogen-fixing crops production. Methane and nitrous oxide are also emitted in large amounts from natural sources. The 100-year global warming potentials of methane and nitrous oxide are recently estimated at 25 and 298 carbon dioxide equivalents, respectively.

Steinfeld estimate that livestock production accounts for 18 percent of anthropogenic GHG emissions expressed as carbon dioxide equivalents. Of this amount, 34 percent is carbon dioxide emission from deforestation, principally in Central and South America, that they assigned to livestock production. However, deforestation associated with livestock production is not an issue in many regions. In the US, the land area occupied by forest increased between 1990 and 2009 and a net increase in forest land area was also reported in Canada.

Of emissions they attribute to livestock production, Steinfeld estimate that globally, methane accounts for 30.2 percent. Like other greenhouse gases, methane contributes to global warming when its atmospheric concentration rises. Although methane emission from agriculture and other anthropogenic sources has contributed substantially to past warming, it is of much less significance for current and recent warming. This is because there has been relatively little increase in atmospheric methane concentration in recent years The anomalous increase in methane concentration in 2007,, has since been attributed principally to anomalous methane flux from natural wetlands, mostly in the tropics, rather than to anthropogenic sources.

Livestock sources (including enteric fermentation and manure) account for about 3.1 percent of US anthropogenic GHG emissions expressed as carbon dioxide equivalents. This EPA estimate is based on methodologies agreed to by the Conference of Parties of the UNFCCC, with 100-year global warming potentials from the IPCC Second Assessment Report used in estimating GHG emissions as carbon dioxide equivalents.

A 2016 study published in Nature Climate Change concludes that climate taxes on meat and milk would simultaneously produce substantial cuts in greenhouse gas emissions and lead to healthier diets. Such taxes would need to be designed with care: exempting and subsidising some food groups, selectively compensating for income loss, and using part of the revenue for health promotion. The study analyzed surcharges of 40% on beef and 20% on milk and their effects on consumption, climate emissions, and distribution. An optimum plan would reduce emissions by 1 billion tonnes per year – similar in amount to those from aviation globally.

High-carbon and Low-carbon Food Choices

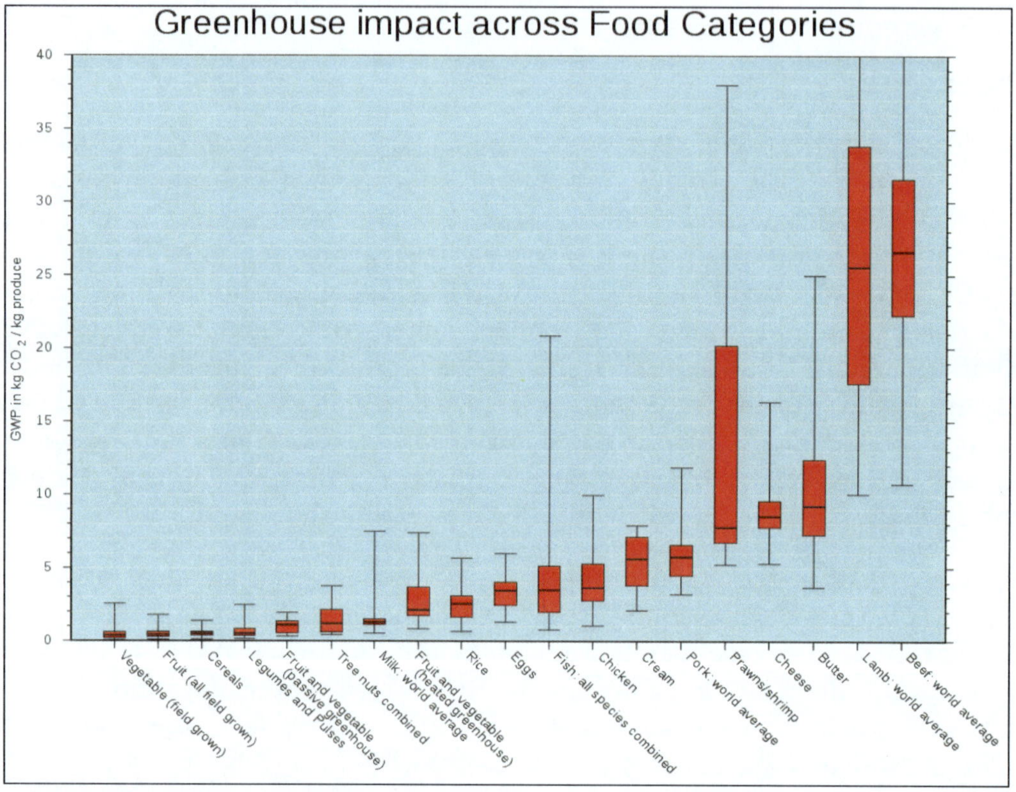

Greenhouse impact across food categories: results from a 2017 review of 389 Life-cycle assessments.

Certain foods require more fossil fuel inputs than others, making it possible to go on a low-carbon diet and reduce one's carbon footprint by choosing foods that need less fossil fuel and therefore emit less carbon dioxide and other greenhouse gases.

In June 2010, a report from United Nations Environment Programme declared that a global shift towards a vegan diet was needed to save the world from hunger, fuel shortages and climate change.

Cundiff and Harris write: "The American Dietetic Association (ADA) and Dietitians of Canada position paper officially recognizes that well-planned vegan and other vegetarian diets are appropriate for infancy and childhood. The American Academy of Pediatrics concurs." However, care is essential to ensure adequate nutrition with a vegetarian diet. For example, there is evidence of higher frequencies of non-anemic iron deficiency among vegetarians versus omnivores and lower zinc status of vegetarian children, in addition to the need for vitamin B12 supplementation with vegan diets. Statistically, a higher risk of dietary deficiency has been reported with vegan diets. Also, some studies indicate higher frequency of hypospadias among children of vegetarian mothers.

China introduced new dietary guidelines in 2016 which aim to cut meat consumption by 50% and thereby reduce greenhouse gas emissions by 1 billion tonnes by 2030.

Industrial versus Pastured Livestock

Beef and dairy cattle can be particularly high in their levels of greenhouse gas emissions. Feed is a significant contributor to emissions from animals raised in Confined Animal Feeding Operations (CAFOs) or factory farms, as corn or soy beans must be fertilised, irrigated, processed into animal feed, packaged and then transported to the CAFO. In 2005, CAFOs accounted for 74% of the world's poultry production, 50% of pork, 43% of beef, and 68% of eggs, Proportions are significantly higher in developed countries, but are growing rapidly in developing countries, where demand is also growing fast. However, in the US, only about 11 percent of soybean acres and 14 percent of corn acres are irrigated; in contrast, about 66 percent of vegetable acres and 79 percent of orchard acres are irrigated. In 1995, commercial fertilizer inputs averaged 11 pounds per acre for US soybean production, versus 157 pounds per acre for US potato production. Soybean meal for livestock feed is commonly produced after extraction of soybean oil (used for cooking, food products, biodiesel, etc., so that only a fraction of processing is assignable to feed. Such examples illustrate that issues relating to irrigation, fertilization and processing for meat production should also be of concern with regard to production of other foods.

In one study, grass-fed cattle were estimated to account for 40% less greenhouse emissions than CAFO cattle However, comparative effects on emissions can vary. in a US study, lower GHG emissions were associated with feedlot-finished beef production than with beef production on pasture and hay. Similarly, a study in New Zealand concluded that environmental emissions per kilogram of beef produced can be reduced by incorporating feedlot finishing in a beef production system. Another factor to be considered is the role of a healthy pastoral ecosystem in carbon sequestration. Rotational grazing of ruminants (cattle, sheep, goats, etc.) and birds (chickens, turkeys, etc.) on untilled pasture land promotes rapid topsoil accumulation, representing a major carbon sink.

Because CAFO production is highly centralised, the transport of animals to slaughter and then to distant retail outlets is a further source of greenhouse gas emissions. However, this can be more or less compensated by reduced transport of feed, where CAFOs are located in feed-producing areas.

In livestock production, emissions are reduced by feeding human-inedible materials that might otherwise by wasted. Elferink state that "Currently, 70 % of the feedstock used in the Dutch feed industry originates from the food processing industry." Among several US examples is the feeding of distillers grains remaining from biofuel production. For the marketing year 2009/2010, the amount of dried distillers grains used as livestock feed (and residual) in the US amounted to 25.0 million tonnes.

Distance Traveled and Method of Transit

Carbon emissions from transport account for 11% of the total carbon emissions of food, of which the transportation from producer to consumer accounts for 4%. However, "food miles" are a very misleading measure; in many cases food imported from the other side of the world may have a lower carbon footprint than a locally produced equivalent, due to differences in farming methods; "local food" campaigns may be motivated by protectionism rather than genuine environmentalism.

When looking at total greenhouse gases (not just carbon dioxide), 83% of emissions come from the actual production of the food because of the methane released by livestock and the nitrous oxide due to fertilizer.

The word locavore describes a person attempting to eat a diet consisting of foods harvested from within a 100-mile radius.

Some studies have criticized the emphasis on local food, claiming that it romanticizes local production, but does not produce very much environmental benefit. Transportation accounts for a relatively small portion of overall energy consumption in food production, and locally produced food may be much more energy intensive than food produced in a better area. Additionally the emphasis on "inefficient" local producers over more efficient ones further away may be damaging.

Processing, Packaging and Waste

Highly processed foods such as granola bars come in individual packaging, demanding high energy inputs and resulting in packaging waste. These products contribute up to a third of total energy inputs for food consumption, as their ingredients are shipped from all over, processed, packaged, trucked to storage, then transported to retail outlets. Bottled water is another example of a highly packaged, wasteful food product. It is estimated that Americans throw away 40 million plastic water bottles every day, and bottled water is often shipped trans-continentally. Carbonated water must be chilled and kept under pressure during storage and transport so as to keep the carbon dioxide dissolved. This factor contributes greater energy usage for products shipped longer distances.

IMPACTS OF CLIMATE CHANGE MITIGATION STRATEGIES ON ANIMAL WELFARE

Climate change is arguably the most important environmental issue of our time. With severe and widespread destructive effects, warming of the planet threatens ecological systems, peoples' livelihoods, and species survival. Animal agriculture is an important source of greenhouse gas (GHG) emissions and has been implicated as a serious contributor to climate change. To reduce the contribution of emissions attributable to animal agriculture, a number of mitigation strategies involving changes to farming practices have been proposed. Although this is an important and timely goal, and many of the proposed solutions seem reasonable on the surface, mitigation strategies can have complex effects on people and animals in practice. While there has been occasional mention, in the global discussion on climate change there has generally been a dearth of attention paid to the animal welfare impacts of the proposed abatement options and some of the suggested livestock management approaches would have severe and wide-ranging impacts on the animals.

At the same time, in other arenas there is a growing international social movement afoot aimed at addressing animal welfare, the physical and psychological state of animals. There are many animal welfare issues associated with commercial farming practices, especially in industrial agricultural production systems, where animals are often confined indoors at high stocking densities. The conditions in which the animals are kept are a matter of serious deliberation in legislative, corporate, investment and trade organizations, among many others around the world. Overarching calls to increase animal productivity, by, for example, pushing animals (either genetically through selective breeding, with hormone drugs, or through feed changes) to produce more milk or meat, or to produce more offspring faster, can also be considered in light of potential health and welfare consequences.

The Earth's surface has been successively warmer over each of the last three decades compared to any preceding decade since 1850. Human activities are responsible (with a 95%–100% probability) for the recent global warming and the marked increase in global atmospheric concentrations of carbon dioxide (CO_2), methane (CH_4) and nitrous oxide (N_2O) to above pre-industrial values. By sector, greenhouse gas pollution originates primarily from industry; agriculture, forestry and other land use (AFOLU); buildings; and transport. Separated out, animal agriculture emits a significant portion, estimated from 7% to 18% of anthropogenic greenhouse gas emissions.

Livestock contribute both directly and indirectly to climate change. Enteric fermentation and manure associated emissions are direct, while production and transport of feed (including the fossil fuels used in manufacturing chemical fertilizers) and land use changes (such as conversion of forest to pasture and crop land) contribute indirectly. About 44% of the emissions generated by livestock are CH_4, which is released during enteric fermentation (eructation in ruminants) and emitted from manure decomposition; 27% are in the form of CO_2 emitted during the production and transport of animal products and feed, and 29% are N_2O attributable to manure and fertilizer. There is a clear need to address the contribution that farming animals makes to global warming.

To combat climate change caused by animal agriculture, a number of mitigation strategies have been proposed, as well as policy options to implement such strategies.

However, by far, the most substantial emissions reductions can come through adaptations in current systems rather than requiring a shift to industrialized systems. This is important from an animal welfare perspective. While there can be negative animal welfare impacts in all production systems, industrialized production has inherently problematic effects on billions of animals globally. Thus, it is important that there is significant alignment in both animal welfare and climate goals.

Mitigation Strategies and Possible Negative Animal Welfare Effects

Feed Changes

Given that feed production accounts for about 47% of livestock emissions, it is a key target for mitigation. However, there are a number of welfare concerns with some of the current proposals to reduce emissions using feed alterations.

Grain Finishing Feedlot Cattle One strategy for mitigating methane emissions with feed changes is by including more concentrates in addition to, or in place of, a proportion of dietary forage. Although emissions per animal may rise as productivity improves, there is generally a reduction in the greenhouse gases released on a per kg-product produced basis. In intensive production systems, grain feeding of livestock often entails removing cattle raised for beef from pastures to provide the finishing ration in a feedlot. This is a common system in the United States and is being emulated in other parts of the world, such as Latin America. In Brazil, the practice of finishing cattle in feedlots increased 50% (from two to three million animals) from 2003 to 2010.

The digestive system of cattle is best adapted to roughage provided by grass-based diets and the natural behavior of cattle is to graze throughout the day. However, the typical routine in feedlots is to gradually reduce the proportion of forage in the ration over time, eventually reaching diets that can be as high as 90% grain. In enlarging Brazilian feedlots, the expected trend is a decrease in the amount of roughage in feedlot diets, with an already observed reduction from 29% in 2009 to 21% in 2011.

In cattle feeding systems with high ratios of concentrate to forages, digestive problems are common. The starches in this concentrated diet are quickly digested and fermentation acids can build up disrupting the normal function of the rumen. Abnormal rumen function and digestive disorders can lead to acidosis, bloat, and if persistent, liver abscesses, and even the foot disorder laminitis Maintaining normal rumen function on these diets is a constant challenge for the feedlot industry; one quarter of cattle mortality in feedlots can be attributed to digestive disorders.

For both acidosis and bloat, increasing the amount of forage in the diet can stimulate chewing and promote saliva production, which have a buffering effect, and thus reduce the likelihood of a sinking pH level and destabilization of the microbial population in the rumen.

Finishing calves in a feedlot usually involves transporting them from their natal pastures. In the United States, nearly two-thirds (63%) of cattle and calves raised for beef are sold through an auction barn after leaving the farm, thus, many calves are moved twice before arrival at a feedlot.

Transport takes a physical and psychological toll on animals. Stressors include unfamiliar surroundings, novelty, noise, vibration, social regrouping, loading and unloading, and feed and water deprivation, although many factors, such as preconditioning and location in the vehicle during the journey can alter the effect. The stress of transport can lead to immunosuppression and an increase in disease susceptibility, including the ubiquitous Bovine Respiratory Disease (RBD) problem.

Concentrated Diets for Pigs In cases where intensification of animal agriculture is accompanied by better access to resources, then there may be benefits for welfare in terms of providing more nutritionally complete rations. However, pigs are also often fed highly concentrated grain-based diets in intensive production. Concentrated diets may be fed to pigs in a finely ground or pelleted ration and are expected to reduce GHG emissions due to their high digestibility. For example, in a case study of intensive pig production in East and South East Asia, the FAO explored changes to swine diets as a potential mitigation strategy, including increased digestibility of feed. In intermediate (not fully industrialized) systems, by changing the feed along with other management changes, emissions could be reduced by 32% to 38%.

Pigs' stomachs are biologically designed for numerous, small meals throughout the day of high fiber feedstuffs. However, in intensive production, they have little access to roughage. Feed processing improves feed-to-gain efficiency, but it is also associated with intestinal problems. These low fiber diets can impact mucosa integrity leaving pigs prone to gastric ulcers in the pars esophageal region of the stomach. This is a problem worldwide. Reports vary, with incidence of pigs showing distinct signs of ulceration between farming operations ranging from zero to nearly 90%. For pigs in the growing/finishing stages of production, ulceration is a common cause of mortality.

The production systems in which most pigs are kept appear to have a large impact on the incidence of these ulcers, as pigs with access to straw, sawdust, or outdoor paddocks have fewer ulcers than those confined on bare, solid or slatted concrete floors.

Feed Additives Feed additives can be used to inhibit CH_4 production in a variety of species. However, some dietary additives that are being tested have potential animal health concerns. For example, *in vitro* studies have shown that fumarate can work as an alternative to CH_4 as a sink for H_2 in the rumen, but feeding fumarate in its free acid form can lower rumen pH and inhibit fiber digestion, while feeding it in salt form can induce toxicity. Effective doses may well exceed current dietary recommendations. Especially in low protein diets, nitrates are also potential CH_4 mitigation agents. However, the intermediate product from nitrate metabolism, nitrite, is toxic, and so feeding it requires a gradual acclimation to a high dietary intake in order to avoid nitrite toxicity. Another proposed supplement, sulfate, has been linked to sulphur-induced polioencephalomalacia, a condition in US feedlot cattle associated with excessive production of H_2S in the rumen from diets high in distillers grains—co-products from the ethanol industry, which are high in sulfate.

Genetic Selection In general, as animals become more productive, greater feed intake is needed to support the energy requirements associated with growth or lactation, but the proportion of total energy requirements required for maintenance decreases. Thus the feed input per unit of product produced is reduced. Further, with increases in productivity, fewer animals are needed to reach a given level of output. In the climate change mitigation literature, this concept is often exemplified with dairy cows. For example, in 1944, the diverse breeds comprising the US dairy cow population produced an annual milk yield of 2074 kg per cow, but by 2007 the genetic pool had shrunk, and Holsteins were the dominant breed, while milk production more than quadrupled to 9193 kg per cow. In 1944 there were 25.6 million dairy cows in the United States with an annual production of 53.0 billion kg of milk, while in 2007 the total US dairy herd had declined to 9.2 million cows, but they produced 84.2 billion kg of milk. Because of this per cow increase in production, the total carbon footprint for the US dairy industry in 1944 was 194 million tons of CO_2-equivalent (CO_2-eq.), while in 2007 the industry's carbon footprint was only 114 million tons. Much of this individual increase in productivity can be attributed to genetic selection, although dietary changes also contributed.

Swine and poultry produce relatively small amounts of enteric CH_4, but their manure can be a significant source of GHG production. Therefore, maintaining or improving feed conversion efficiency in these species, and subsequently reducing the volume of manure produced for a given level of output, is considered a major strategy for mitigating CH_4 and N_2O. However, there is evidence across species that overemphasis on productivity in genetic selection can lead to unintended negative effects on animal welfare.

Effects of Breeding for Yield on Dairy Cow Welfare While selective breeding has been successful in producing cows that yield more milk with less feed, this change is not without costs. Detrimental, unintended side-effects on the health of dairy cows have accompanied the physiological demands of greater productivity.

Following the birth of her calf, a dairy cow enters a state of negative energy balance. Genetic selection for increased milk yield increases feed intake, but high yielding cows can experience greater negative energy balance than cows producing less, so they are vulnerable to health problems.

There is a strong correlation between maximum metabolic load and laminitis, lameness and ketosis. Failure to meet the nutritional demands of highly productive dairy cows can also lead to an array of additional metabolic diseases including fatty liver disease, ketosis and periparturient hypocalcemic paresis (milk fever).

Mastitis, a bacterial infection of the udder, is the most commonly reported health problem in the US dairy industry, affecting 16.5% of cows. The udder and teats of mastitic cows are sensitive to the touch, and in severe cases, clinical signs include swelling and pain sometimes accompanied by fever and depression. It is one of the leading reasons for culling of cows. Although there are many risk factors for mastitis, such as lack of cleanliness in cow housing areas, research has consistently shown an association between mastitis and the amount of milk produced. For example, a study of over 3000 cows in 85 trials in eight different countries found that even with similar management conditions, higher yielding cows contracted more mastitis than lower yielding cows. Many studies have reported that milk yield and mastitis are genetically correlated.

Further, high genetic merit dairy cows require high energy feeds to meet the substantial energy demands of early lactation. They often receive a high proportion of cereal grains, and one model found that cows selected for greater milk fat and protein fed a 50% grain diet had lower emissions (1.05 kg CO_2-eq. per kilogram of energy-corrected milk), than cows with a 25% grain diet and summer grazing (1.19 kg per kilogram of energy-corrected milk). However, dairy cows are prone to digestive disorders such as sub-acute ruminal acidosis (SARA). Diets that are rich in grains alter rumen metabolic balance and can induce an explosive proliferation of *Streptococcus bovis* and *Lactobacillus* bacteria species. The resulting rumen acidosis eventually leads to erosion and ulceration of the ruminal epithelium and chronic symptoms including lethargy, reduced feed intake, diarrhea, and discomfort among others.

Effects of Breeding for Productivity on Swine Welfare Increases in the productivity of swine, largely achieved through genetic selection, have been cited as climate change mitigation progress. For example, a 2013 FAO report cited data provided by the US National Pork Board. Between 1959 and 2009, there was a 29% increase in the number of hogs marketed in the United States, yet the size of the breeding herd has been reduced by 39%. This is due in part to the steady increase in litter size from 7.10 in 1974 to 9.97 piglets in 2011 and the concomitant change in the amount of pork produced from a single breeding animal from 775 to 1828 kg. Feed efficiency also increased by 33%. This has reduced the carbon footprint per 454 kg of hot dressed carcass weight produced by 35%. However, the original source for this calculation does not appear to be peer-reviewed, and total emissions in the US pork industry have increased overall with greater pork production. Despite this, the suggestion is that the type of throughput achieved in the United States, if adopted more widely in the developing world, could be an approach to climate change mitigation. However, some of the key unintended animal welfare consequences that may accompany such a change include low piglet viability and severe feed deprivation of breeding sows in order to prevent them from becoming obese.

Selectively breeding for increased litter size reduces the proportion of surviving piglets. Herpin postulated that genetic changes have altered body fat metabolism, body composition, and hormonal state, resulting in lean tissue growth that makes piglets heavier but less mature at birth compared to lines that have not been so intensely selected for production traits. This reduces piglet survival rate, which is an intractable problem in the pork industry. Selection for leanness may have also inadvertently decreased the nutritional quality of sows' milk, in turn further affecting survival of the piglets.

Selectively bred for productivity, growing/finishing pigs in the United States are fed to satiety to meet their full genetic potential, but breeding animals must often be feed restricted to prevent excessive weight gain and fat deposition. Confined sows are typically fed just half the amount they would choose to consume if fed *ad libitum*. This leads to a chronic hunger problem including persistent, unfulfilled feeding motivation. Short feeding times and lack of an outlet for natural foraging behavior are underlying causes of abnormal stereotypic behavior.

Effects of Breeding for Productivity on the Welfare of Chickens As with swine, genetically linked productivity improvements in poultry have also been tied to climate change benefits through feed conversion efficiency and the extremely rapid growth rate of modern broiler chickens (raised for meat). Broilers now grow from hatch to market weight in five weeks or less. Annual growth rates have increased 3.3% over the past 50 years, and broiler chickens now grow more than four times faster than they did in 1957. Using a life-cycle assessment model, one study estimated that genetic changes in average U.K. broiler chickens have achieved a 1% reduction per year in greenhouse gases over the last 20 years.

These animals now grow so fast that muscle outpaces bone development, leading to metabolic bone disease. As a result, broiler chickens often suffer from leg deformities and lameness, including tibial dyschondroplasia, angular bone deformities, spondylolisthesis, epiphyseitis, and in severe cases, rupture of the gastrocnemius tendon. Studies are consistent in showing that approximately 26%–30% of broiler chickens by 40–42 days of age suffer from gait defects severe enough to impair walking ability. Additional research strongly suggests that while conformational differences account for some gait differences, birds at this level of lameness are probably in pain.

Not only do rapidly growing broiler chickens suffer from musculoskeletal diseases, but they are also afflicted by metabolic disorders. Ascites is a condition in which rapidly growing broiler chickens do not have the heart and lung capacity needed to distribute oxygen throughout the body and is a leading cause of mortality as the birds reach market weight. Sudden Death Syndrome (SDS) is associated with acute heart failure caused by dysrhythmias. Young birds die from SDS after sudden convulsions and wing-beating, and are frequently found lying on their backs. The condition has been recognized since the 1950s as more broiler chickens were grown in large numbers for commercial production. Sudden death syndrome (SDS) and ascites together can account for 50% of the mortality of highly productive broiler chicken strains.

The parent generation of broiler chickens grown for meat shares the same genetics as their progeny. Like breeding sows, these broiler breeders must be feed restricted, or they will suffer from health and reproductive disorders due to weight gain. Parent birds are usually feed-restricted starting when they are as young as one week old. In many parts of the world broiler breeders may be fed on a "skip-a-day" regimen, in which the animals are fed as infrequently as every other day—though this practice has been outlawed in several European countries. Experimental studies suggest that artificial selection for increased body weight may have altered the brain mechanism controlling satiety and appetite, and evidence from behavioral studies suggests that feed restriction interferes with learning and causes stress, boredom, and chronic hunger. Breeding birds receive only

25%–50% of the amount of feed they would otherwise eat if given free access. While free-range chickens normally devote about 50% of their daily time budget to foraging, feed restricted parent birds can consume their daily feed allotment in as little as 15 min. Feed restriction is believed to cause undernourishment, nutritional deficiency, and abnormal behavior including increased pecking at non-feed objects, pacing, and heightened aggression. Given that target body weights for broiler breeders have changed little in the past 30 years, but broiler body weight continues to increase, the welfare of parent birds may become more serious in the future as they will be feed deprived to an even greater extent.

Genetic Selection with Regard to Welfare While genetic selection for productivity can have negative side-effects for animal welfare, this is not necessarily always the case. Although progress may be slow, it is possible to select for both welfare associated traits and productivity simultaneously. For example, while there is an overall negative correlation between walking ability and growth rate of broiler chickens, some individuals do have both good gait scores and high productivity. Given enough time, it might be possible to successfully select for both traits. Where animals can be bred to produce the same amount of meat or milk, but with less feed, GHG emission intensity (CH_4/unit animal product) is reduced. The most efficient animals produce less CH_4 and N_2O than the population average, and there might not be a further welfare impact if selection focused on these animals. Direct selection on reduced CH_4 production could also be a possibility. For example, in one study of sheep the heritability of g CH_4/day was 0.29 with no genetic correlation with live weight, and a slightly economically favorable correlation with fleece weight.

Mismatching Genetics and Environment The single most effective GHG mitigation strategy in many regions of the world is to increase animal productivity. Raising imported animals selected for production traits, instead of traditional breeds is part of that strategy. However, the FAO also cautions that while achieving the genetic potential of highly productive breeds is important, these animals should not be imported into environments where climate or management limitations would prevent the animals from achieving this potential. In new regions or countries, animals from genetic lines with high production potential may actually perform worse than native breeds or crossbreeds, especially compared to those well-adapted to local disease and climactic conditions. Some regions have seasonal fluctuations in feed quality and nutrient management, and, as high genetic merit breeds have exacting nutritional requirements, they do not always thrive as expected.

In Africa, resistance to trypanosomiasis is a factor that limits the geographical distribution of cattle breeds. Indigenous taurine (humpless) breeds of cattle including the N'Dama and West African Shorthorn are genetically trypanotolerant, but European taurine breeds have no resistance, and even Zebu cattle imported from other regions of the continent do not thrive in the humid tropics of West Africa occupied by tsetse flies, vectors of trypanosomiasis. Clearly there are limits to importing more highly productive breeds in this large swath of the world, and such circumstances call in to question the notion of "genetic improvement"—indeed the improved breeds here would appear to be the ones that can best survive the local conditions.

In Brazil, more fertile and adapted breeds, such as the Brazilian Gir, are increasing in popularity. Indigenous breeds are better adapted to local conditions, and more readily able to avoid predators, tolerate heat or other weather extremes, resist parasites and endemic diseases, and scavenge for feed or consume local feed resources. For example, in the subtropical climate zone, with seasonal rainfall patterns that do not permit continuous vegetation growth, Zebu and Sanga cattle are more resistant to ticks and tick-transmitted diseases as well as feed scarcity and seasonal high temperatures.

While promoting highly productive genetics to address climate change will reduce GHG emissions in some ways, maintaining biodiversity and the genetic resources of locally adapted breeds will be needed to complement other climate change mitigation strategies and to ensure the animals thrive. A potential solution may be more research and genetic selection of local breeds, but only with due consideration to welfare. Cradle-to-grave analyses that include the full range of trade-offs will become increasingly necessary. Genetic suitability for survival in less hospitable regions with more variable climates and diverse, changing feed resources is currently, and may become even more so, an essential requirement for both human and animal welfare.

Growth and Yield Promoting Biotechnologies Some authors are calling attention to the wide application of growth promoting technologies, including β-adrenergic agonists (βAAs) and recombinant bovine somatotrophin (rbST), for their climate change mitigation potential. While the research to date has focused on US production systems, the call is being made in the context of meeting increasing global demand for animal protein. Zilpaterol hydorchloride and ractopamine hydrochloride are βAAs that are used as feed additives in grain rations for cattle. βAAs are fed in a 20–40 day period prior to slaughter to improve feed efficiency and weight gain. The growth hormone rbST is used in the dairy industry to increase milk yield and feed conversion efficiency. However, both rbST and βAAs are cause for substantial animal welfare concern.

β-Adrenergic Agonists Using Life Cycle Analysis (LCA) modeling, Stackhouse found that the use of βAAs, in combination with antibiotics, ionophores, and hormone implants, as is common in US feedlots, decreased the carbon footprint of Angus beef production by 9%, and therefore advocated their use as a potential mitigation option. Cooprider calculated that feedlot cattle fed ractopamine hydrochloride along with hormone implants and antibiotics had a 31% decrease in emissions per finished steer given the improved feed efficiency. However, anecdotal reports of changes in behavior and more cattle deaths when βAAs were being fed has led researchers to investigate. One recent study found that cattle fed βAAs along with a growth implant (trenbolone acetate and estradiol) engaged in more agonistic behavior including bulling and pushing. A large-scale investigation by two major US agricultural universities examined datasets provided by feedlot operators also confirmed the field observations, finding that the cumulative death rate is as much as 90% greater in animals administered a βAA, and that 40%–50% of deaths among groups administered a βAA could be explained by administration of the drug. Imports of animal products containing beta agonists is banned in many countries, including the EU, Russia, and China.

The βAA ractopamine hydrochloride is also fed to swine, where the side-effects that have been elucidated to date include elevated heart rate, difficulty walking, increased aggression, abnormal behavior, and hoof lesions. Observations at slaughter plants additionally find that difficulty walking due to ractopamine may contribute to a greater incidence of non-ambulatory (or "downed") pigs, those too weak to stand and walk on their own accord. The number of studies investigating the animal welfare effects of βAAs is, thus far, small, but given the problems identified already, calls for their increased use clearly should be made more cautiously.

Recombinant Bovine Somatotrophin Also called Bovine Growth Hormone, rbST is a protein hormone that redirects nutrient partitioning during lactation and can increase milk yield by 8%–36%. On a per unit of milk produced basis, it has been show that rbST reduces the maintenance energy and protein requirements of cows by 11.8% and 7.5%, respectively, and total feed requirements by 8.1%. It also reduces N excretion on a per liter of milk produced basis. Although overall N excretion increases in rbST treated cows, because they consume more feed, each cow produces more milk, so estimates calculated on a per liter basis report a reduction in nitrogen excretion between approximately 9% and 15%. Per unit of milk, the dilution of maintenance conferred by the use of rbST can result in a reduction in manure production by 6.8% and CH_4 output by 7.3%.

While rbST is banned in the European Union, Canada, Japan, Australia, and New Zealand, the drug is approved in the United States, despite substantial controversy. One of the primary welfare concerns associated with the increased milk production caused by rbST is the potential risk for mastitis. Some of the initial studies on rbST indicated that its use was linked to an increase A 1999 report of the Scientific Committee on Animal Health and Animal Welfare (SCAHAW), a working party of independent scientists appointed by the European Commission to assess rbST, reviewed approximately two dozen studies and concluded that rbST usage increases the risk of clinical mastitis and the length of infections. This finding was supported by a review published by the Canadian Veterinary Medical Association (CVMA). The CVMA performed a series of meta-analyses to combine data from all randomized clinical trials that had been published in peer-reviewed journals or which were provided by Health Canada for the registration of rbST in the country. The CVMA found that rbST increased the risk of clinical mastitis by approximately 25%, closely agreeing with the SCAHAW report. Subsequently, rbST was not approved for use in Canada, due to animal health concerns.

The debate between researchers in different countries largely centers on the question of whether increases in mastitis are due to rbST treatment *per se* or to the increase in productivity caused by rbST, the same as would be expected when genetic selection leads to greater milk production. White argue that treatment differences in mastitis are within the range predicted using estimates from genetic studies. And using this logic they conclude that rbST treatment had no effect on mastitis. In contrast, SCAHAW argue that what matters is the result on the welfare of the animals, which is the same whether it is the drug itself or the increase in productivity that causes the mastitis.

In addition to problems with mastitis, there is also concern that rbST treatment may increase the risk of lameness and foot and leg disorders. While many early studies failed to find a detectable difference between rbST treated and control animals. others found a slight increase in the number of cows culled due to lameness and slightly more feet and leg problems. As with the early work on mastitis and rbST, in studies with small cow numbers, low statistical power may have prevented detection of effects, and slight increases would be hard to find.

However, both the SCAHAW committee and the CVMA review panel found an increased incidence of foot and leg disorders associated with the long-term administration of rbST. The CVMA review panel found that cows treated with rbST were at an estimated 55% increase in the risk of developing clinical signs of lameness, and this is now reflected on the Posilac drug label, which states “Cows injected with Posilac may have more enlarged hocks and disorders of the foot region”.

Recombinant bovine Somatotropin has also been shown to reduce the ability of cows to cope with high environmental temperatures, which can lead to heat stress. It is, thus, not advisable in geographic regions where severe heat can be expected, especially if management options to alleviate heat stress are insufficient.

Clearly the use of rbST carries unintended risks to the welfare of animals, and these should be seriously considered before such growth-promoting compounds are touted as a possible mitigation strategy for climate change.

Species Shifts

In addition to increasing productivity within species, the current global trend toward intensively producing greater numbers of non-ruminant animals—pigs and chickens—is often viewed as a more efficient form of meat production compared to traditional ruminant cattle, sheep and goat rearing systems. To meet the growing demand for animal protein in emerging economies, the expansion in supply is largely coming from increased local production, as opposed to from imports and in these regions, livestock systems are rapidly shifting toward monogastric species. Pigs and poultry account for 77% of the expansion in production, a trend that is often accompanied by intensification, vertical integration, geographic concentration, up-scaling of production, and feeding more grain- or concentrate-based diets—in short, industrialized farming.

Monogastrics have improved GHG efficiencies compared to cattle. Cattle accounted for 77% of total non-CO_2 GHG emissions from the livestock sector in 2000, while monogastrics accounted for just 10%. It takes at least three times as much land to produce one kilogram of beef as it does to produce one kilogram of pork or chicken. Cattle contribute about seven times the CO_2-eq. emissions of pigs and chickens. This biological difference can be explained by three main factors: greater feed efficiency, lower enteric CH_4 emissions, and faster rates of reproduction. CH_4 emissions from ruminants originate from both manure and enteric fermentation, while CH_4 emissions from monogastrics are primarily associated with manure alone. Because sows and hens have a relatively large number of progeny and earlier sexual maturity, the number of breeding adults per animal slaughtered for human food is reduced in these production systems. Together, these factors account for the reduction in inputs and lower emissions of pig and poultry production worldwide.

If the increasing consumption of animal protein in developing countries will be supplied in a substantial way by expansion of pork and chicken industries, and if the expansion will be in the form of specialized, intensive confinement facilities that typify the bulk of production in North America and other developed regions, then there is a serious potential for a concomitant increase in the animal welfare problems that characterize those systems.

The intensive production of modern strains of poultry is the cause of one of the most severe and wide-spread animal welfare problems in the world. Globally, nearly 60 billion chickens are slaughtered for meat each year and given the wide prevalence of metabolic problems and painful leg disorders, the intensive production of meat chickens already constitutes one of the most problematic aspects of intensive agriculture. Sweeping calls to use *more* of these compromised animals to meet efficiency goals is alarming.

In an ethical analysis, overall welfare would be a function of both the state of the individual animal and the number of animals subjected to that level of welfare. It takes far more animals to produce the same amount of meat when beef is replaced in the diet with chicken, compounding the overall magnitude of any negative animal welfare impacts.

Another of the most salient issues with intensive confinement of animals in agriculture is the way those animals are housed. Intensive systems for pork and poultry production remove animals from the land and instead raise them in indoor, total confinement facilities. While well-managed, large-scale production can have some animal welfare advantages, such as access to resources that might be economically out of reach to smaller producers, they often greatly limit the space allotment per animal. In the case of breeding sows in the pork production industry the animals are further partitioned into enclosures that prevent normal movement and severely restrict the animals' natural behavior, which is a serious welfare problem.

Industrial Production of Pigs Swine are highly intelligent, social animals with complex behavioral repertoires. Yet, in many countries, intensive production facilities house female breeding pigs in narrow gestation crates, metal stalls that measure approximately 0.61 m by 2.31 m, which is only slightly larger than the sow's own body. Each sow can take a step forward and backward, but she cannot turn around for the length of her pregnancy, which is approximately 114 days. Due to decreased fitness from chronic lack of exercise, crated sows have lower bone strength and muscle weight, and have higher resting heart rates, compared to those permitted more freedom of movement and exercise.

Fed concentrated grain-based diets that can be consumed quickly, and without opportunity to express species-typical foraging behavior, sows in intensive confinement operations often begin to display abnormal, stereotypic behavior including repetitive bar-biting, head-weaving, sham-chewing and drinker pressing. Rooted in stress and lack of stimulation, these "stereotypies", are common in captive animals kept in barren, restrictive environments. Stereotypic behavior is often reduced when sows are fed a high fiber, bulky diet instead of, or in addition to, concentrates, although some fiber sources, such as soybean hulls, may be less effective. Ethologists attribute this stereotypic behavior to boredom and frustration resulting from an impoverished environment, confinement, restraint, and unfulfilled behavioral needs.

Most large, industrialized confinement facilities for pigs utilize a liquid manure handling system with slatted floors. While a 2013 FAO review noted that slatted floor manure systems tend to decrease GHG and NH_3 emissions compared with deep litter systems, these bare, concrete floors are hard on the animals' feet. One survey found that pigs allowed outdoor access had a lower prevalence of foot and limb injuries, while those confined indoors on hard, slatted flooring had more bruising, calluses, locomotion problems, and adventitious bursae. The problem is especially acute for piglets; foot and limb injuries affect 60% to 90% of young piglets housed indoors, while only about 9% of piglets housed outdoors suffer these types of injuries. While outdoor production does have its own suite of welfare concerns, including severe weather and predators, in farrowing huts common to pasture-based systems, soil covered with straw provides a soft, non-abrasive, protective surface. Behavioral testing has shown that pigs prefer earthen floors over concrete. While bedding, such as straw provides comfort, warmth and outlets for natural rooting and nesting behavior, it is not usually provided in indoor operations due to cost, difficulty of cleaning, and incompatibility with slatted floors.

There are however, more welfare-friendly intensive systems. For example, sows can successfully be housed in group pens as opposed to individual gestation crates, as long as they are managed well to prevent aggression. Group housing systems work best if they deliver the feed in a non-competitive way, such as by using trickle feeding, feeding stalls, or Electronic Sow Feeding systems (which use a computer to precisely deliver each sow's daily allotment). Straw bedding can be incorporated into these systems in purpose-built un-slatted sections of the barn. However, N_2O emissions in a straw-based, deep litter group housing systems are usually greater compared to concrete slatted floors.

Further Disadvantages of Shifting Production Systems Another problem with large, industrialized animal production sites is the ratio of animals to human caretakers. With increased farm size, new technologies and increased mechanization, such as automated feed and water delivery and manure removal, coupled with economic pressure to reduce labor, the result is that fewer workers now tend to more animals. As such, individualized attention to each animal is generally lacking. With the use of efficiencies in pen and barn design, one person may now be responsible for the care of 8000 pigs per day. One of the main daily tasks on a commercial-scale broiler grow-out facility is picking up the dead birds. While large farms can be well managed to reduce death losses, the health of the animals on large operations is often addressed at the herd or flock level, and individual veterinary care is not generally provided. An individual animal who becomes injured or ill is likely to go unnoticed, and could experience prolonged pain and suffering before their eventual death.

On top of the animal welfare concerns, the industrialization of agriculture carries with it numerous other societal implications. Switching to large-scale use of monogastrics threatens genetic diversity, and subsequent genetic adaptation of domestic species to climate change. It could also mean the proliferation of more and more serious environmental issues, including air, soil, and water pollution. For example, the rapid expansion of pig production in the coastal areas of China, Vietnam, and Thailand is thought to be a major source of nutrient pollution in the South China Sea. Runoff is causing red tides and threatening fragile mangroves and coral reefs.

System changes, however, (e.g., from extensive to industrialized production) are not required for some of the most substantial emissions reductions, which can be achieved within production systems, across all species and regions. The FAO found that applying the techniques of 10% of the most efficient producers to current systems, by for example, improved pasture management and feed quality, and better animal health, among others could cut nearly a third of emissions from the farm animal sector. Production system changes further reduced emissions a mere 2%. Clearly, there is a large opportunity to improve emissions by picking the low-hanging fruit, *i.e.*, the least efficient producers, while achieving multiple benefits, such as animal welfare, or at least avoiding additional negative impacts. Such improvements should also aim at improved food security (not just increased production) and fit well within the concept of "sustainable intensification".

Potential Solutions: Mitigation Strategies that Both Improve Animal Welfare and Reduce Environmental Impacts

While it is clear that animal welfare concerns for many different climate change mitigation strategies exist, fortunately a great many more potential solutions have benign or even beneficial welfare outcomes. Many of these options are strategies for extensive production, systems that, if improved, have very high welfare potential. Basic improvements to these systems would also have the co-benefit of improving the livelihoods of smallholder farmers. With more attention paid to the health, physiology, and behavioral effects on animals at the outset, triple-win strategies can be championed that mitigate climate change, improve food security, and improve animal welfare.

Improved Animal Health and Longevity

Where climate change solutions and improved animal welfare may dovetail best begins with a focus on the animals. Improving the efficiency of animal production systems by improving animal health and reducing mortality could reduce CH_4 and N_2O emissions considerably by reducing losses due to health related production declines and emissions attributed to animals that die before they can reproduce or produce consumable products. In locations where basic veterinary care, vaccinations, clean drinking water, shade, husbandry skills, technical knowledge, financial aid and other vital resources are lacking, there is great potential for improvement. For example, in many developing African and Asian countries, approximately 80% of poultry is still located in village flocks. Globally, backyard flocks kept for meat and eggs contribute 42.5 million tons of CO_2-eq. in greenhouse gas emissions and have higher emission intensity than intensively confined chickens, due in part to the low feed quality of scavenging birds and the toll that disease and predation take on productivity. Vaccinations, providing simple, low-cost protection for newly hatched chicks from predators, supplementary feeding of balanced rations, and basic biosecurity education could simultaneously improve the welfare of the flocks, alleviate poverty, and potentially reduce greenhouse gas emissions. The climate change mitigation potential of these improvements has not specifically been studied, to our knowledge.

Extending dairy cow lifetime is another potential improvement. Cows can easily live to 15 years of age or longer, but on most intensive, indoor dairy production facilities that typify developed countries the lifespan of a cow is typically closer to six years. A USDA summary reports the proportion of dairy cows permanently removed from the herd for reproductive problems was 26.3%; 23.0% were removed for udder or mastitis problems, 16.0% for lameness or injury, and 16.1% for poor production. Similar factors for culling are reported for Canada and European countries. In one study of 14 Irish dairy herds, the rate at which cows were replaced in the herd increased from 16% in 1990 to 27% in 2003, an increase of 0.8% per year, and the proportion of cows culled for infertility increased 2.2% per year, from 27% in 1990 to 40% in 2002. Improved longevity would reduce the total lifetime emissions of dairy cows when accounting for the resources needed for rearing replacement heifers. The proportion of total methane emissions produced by replacement heifers has been estimated to be up to 27% for U.K. dairy herds. Relative to the period from birth to first calving (a period that usually lasts from 25 months to about 3 years, depending on the region), a greater number of lactation periods per lifetime will reduce overall CH_4 loss per unit of milk yield. By improving the lifespan of a dairy cow from 3.02 to 3.5 lactations, methane emissions would be reduced by 3%. Selection for increased productive life could also lead to a decrease in disease incidence of dairy cows.

In the future, one possibility for addressing the lack of individual attention on industrialized farms could be "Precision Livestock farming" (PLF). In PLF, sensors continuously monitor key indicators, such as animal movement, feed intake, or temperature and alert caretakers when a problem arises. Examples include using cameras to monitor the behavior of hens, pedometers to monitor estrus behavior of dairy cows, and automated scales for measuring weight gain of chickens. PLF has good potential for automated detection of animal health problems. For example, auditory sensors can be used to detect disease in fattening pigs by analysis of coughing sounds, which change qualitatively when the respiratory system is infected. However, it is uncertain how applicable these highly technical solutions would be in developing countries, and there is a danger of losing genuine husbandry skills and knowledge when relying even more heavily on automation.

Improved Animal Nutrition

Changes to animal diets can reduce greenhouse gas emissions and increase animal productivity, but they can also risk harming animal welfare. However, bolstering nutrition by supplementing poor diets or using improved forages can both reduce greenhouse gas emissions and improve welfare at the same time. In regions of the world with a seasonally dry tropical climate, such as Africa, the low nutritional value of most animal feeds during the dry season is a major constraint on animal productivity. In developing countries, ruminant animals are often fed on low quality grasses and crop residues (stover). Inadequate nutrition is a key factor limiting fertility. A large potential to mitigate emissions exists in these low-yield systems. For example, modeling of small ruminant production in West Africa suggests that improvements in forage digestibility, along with other animal health and husbandry measures, can potentially reduce emissions by 27% to 41% of baseline emissions, amounting to 7.7 to 12 million tons of CO_2-eq.

Improved forages can be used as a climate change mitigation strategy in both developing and developed regions. Legumes available for grazing ruminants in the United States include clover, alfalfa, crown vetch, and birdsfoot trefoil (lotus). These nutritionally valuable, high protein forages can be incorporated into established pasture or rangeland with interseeding. Not only can improved pastures reduce CH_4 emissions from animals, but carbon sequestration in grassland soils can also be increased by planting nitrogen-fixing legumes.

The mitigation potential of feeding legumes has been the subject of a substantial number of studies. Simulation results in one study showed that CH_4 production is 28% lower with legume forage (alfalfa hay) as compared to grass (timothy hay) when expressed as a proportion of gross energy (GE) intake, and 21% lower when expressed relative to digestible energy (DE). These results are corroborated by multiple field studies, which also find lower CH_4 with pasture legumes for sheep and cattle. For example, Waghorn found that when fed to sheep, a sulla/lucerne feed mixture produced just 19.0 g CH_4 per kg DMI compared to fresh pasture, which produced 25.7 g CH_4 per kg DMI. In a second trial, lotus produced just 11.5 g CH_4 per kg DMI, and by using polyethylene glycol binding to remove condensed tannins (CT), the researchers demonstrated that CT played an important role in this reduction. Condensed tannins not only lower CH_4 production in sheep, but may also have other beneficial effects including improving wool growth and reducing intestinal worm burdens. However, at least one study failed to find reduced methane production feeding white clover to sheep and so more study is required.

One caution: legume consumption in some cases is associated with animal health issues, such as fetal hydrops in periparturient ewes and bloat with some tropical legumes. Feeding legumes must, therefore, be carefully managed to avoid animal health problems.

Many other feeding management strategies that can reduce greenhouse gas emissions without compromising animal welfare could be better emphasized and further explored. These include improving silage quality with more effective preservation methods ; adding dietary fats, such as coconut, linseed, soybean, or sunflower seed oils; changes to pasture management such as matching grazing intensity and pasture quality supplementation of the diet with enzymes or yeasts, and precision feeding (better agreement between the supply of N in the diet and requirements of the animal, in order to reduce N excretion), to name just a few.

Improved Manure and Land Management

In addition to improving animal health, nutrition or longevity, a great many more climate change mitigation options remain that could be used without negatively affecting the welfare of animals. Ruminants may be key parts to well-managed, mixed crop-livestock systems, helping to reduce a number of environmental impact. Well-managed grazing can improve soil organic carbon and nitrogen content and partially offset net GHG emissions. Most manure management options are probably welfare-neutral, unless collection systems require the intensive confinement of animals. Hristov provide a thorough summary of such options including reducing manure storage time, composting, and carefully timed land application. Several techniques for manure application also hold mitigation promise, for example subsurface application, application during non-rainy periods, and controlling soil pH. Perennial crops can be used to improve soil structure, water retention capacity and, in turn, carbon sequestration capacity. Converting from conventional tilling to conservation or no-till cropping practices, especially with the simultaneous use of cover crops, can reduce soil disturbance and improve soil biodiversity and is another way to sequester carbon. Other options include restoration of degraded range land (re-vegetation, water conservation), and silvopastoral and agroforestry systems among many others.

Reduced Animal Numbers

The demand for meat and dairy products is rapidly growing. This demand is attributable to the growing world population, urbanization and increasing affluence in emerging economies. The projected global growth in annual consumption of meat and milk is 5.41% and 3.43%, respectively. In Asia this growth is projected to be even greater, at 7.99% and 11.85%. Meat consumption is expected to continue to grow in developing countries until they meet the levels of developed countries, with poultry accounting for around 50% of the increase. If this demand is met, the growing number of animals will substantially increase the problem of climate change, even if production efficiencies per unit of product produced are realized.

The growing demand for animal-based protein is usually presented as inevitable. While the general trajectory seems certain, the call to fill this demand in every case is not necessarily prudent. Reducing animal numbers could come, in part, with changes in production efficiency. However, these production-side efficiency approaches seem, in some ways, to avoid the heart of the problem.

Feeding animals grain to produce animal-based food is inherently inefficient. For every kg of chicken live weight gain, 1.7 kg is required in feed, and chicken is the most highly feed efficient species; the ratio of feed to gain for pigs is 2.4 to 1 and for cattle it is as high as 10 to 1. The production of feed for animals can have a larger environmental impact in terms of energy use and global warming, than the management of those same animals and their manure on the farm. Using IPCC figures, Garnett estimates that the reductions achievable for agriculture mitigation options amount to only about 30% of the total impact ; following this she argues: "hence the conclusion that we need also to moderate our consumption of livestock products seems inescapable".

One approach to this is to address the immense amount of food that is wasted. This category is broad, including waste at harvest, storage, processing, and a number of post-consumer pathways. The inefficiencies above suggest that waste of animal-based foods carries a particularly high environmental cost. The FAO has estimated global food waste at about 1/3, with waste of meat products of about 20%. Smith found savings of about 6% in food crop area, and 13%–47% GHG savings in 2050 through waste reduction compared to the reference case.

Another approach focuses specifically on consumption, particularly in developed countries and in populations that face chronic disease due to overconsumption. Many studies show that reducing consumption of animal products would have a substantial effect on reducing GHG emissions, and the IPCC recognized this potential in their Fifth Assessment Report. Smith found greater mitigation potential in the Agriculture, Forestry, and Other Land Use (AFOLU) sector on the consumption rather than production side, when incorporating associated benefits in spared land use (e.g., afforestation and bioenergy).

While variables such as energy used in transport, processing and refrigeration do not permit blanket statements about animal products over alternatives, meat, seafood, and some dairy products tend to have large GHG emissions per kg of product at the checkout as compared to most plant-based foods. Some analyses show that vegetarian and vegan diets have potential GHG savings of 22% and 29% in the UK and US, respectively. However, a 2014 study conducted in the UK found that dietary GHG emissions for meat-eaters were approximately twice as high as those of vegans.

Vegan and vegetarian diets also have lower environmental impact including less land use, water, and fossil fuel requirements. Reducing climate change mitigation costs, as well as potential to improve public health, have also been cited as a co-beneficial effects.

Substantial effects could be realized even by simply reducing meat consumption, such as through efforts like the Meatless Monday campaign, or partial substitution of plant protein in ground and processed meats. Another potential future alternative is cultured meat production. Cultured meat is laboratory produced using *in vitro* tissue engineering techniques, and initial LCA estimates show that it could lead to 78%–96% lower GHG emissions than conventionally produced European meat, while simultaneously avoiding all of the animal welfare impacts of intensive production.

References

- Constanze Werner et al. (2018): Biogeochemical potential of biomass pyrolysis systems for limiting global warming to 1.5° C.Environmental Research Letters, 13(4), 044036. Doi:10.1088/1748-9326/aabb0e
- Science-environment-26980837, news: bbc.com, Retrieved 7 January, 2019
- "Geoengineering the climate: science, governance and uncertainty". The Royal Society. 2009. Retrieved September 10, 2011
- "Intergovernmental Panel on Climate Change (IPCC) Global Surface Warming Scenarios", Multimedia Atlas of Global Warming and Climatology, SAGE Publications Ltd, 2014, doi:10.4135/9781483351384.n48, ISBN 9781483351384
- "Position of the American Dietetic Association and Dietitians of Canada: Vegetarian diets". Journal of the American Dietetic Association. 103 (6): 748–765. 2003. Doi:10.1053/jada.2003.50142. PMID 12778049

Permissions

All chapters in this book are published with permission under the Creative Commons Attribution Share Alike License or equivalent. Every chapter published in this book has been scrutinized by our experts. Their significance has been extensively debated. The topics covered herein carry significant information for a comprehensive understanding. They may even be implemented as practical applications or may be referred to as a beginning point for further studies.

We would like to thank the editorial team for lending their expertise to make the book truly unique. They have played a crucial role in the development of this book. Without their invaluable contributions this book wouldn't have been possible. They have made vital efforts to compile up to date information on the varied aspects of this subject to make this book a valuable addition to the collection of many professionals and students.

This book was conceptualized with the vision of imparting up-to-date and integrated information in this field. To ensure the same, a matchless editorial board was set up. Every individual on the board went through rigorous rounds of assessment to prove their worth. After which they invested a large part of their time researching and compiling the most relevant data for our readers.

The editorial board has been involved in producing this book since its inception. They have spent rigorous hours researching and exploring the diverse topics which have resulted in the successful publishing of this book. They have passed on their knowledge of decades through this book. To expedite this challenging task, the publisher supported the team at every step. A small team of assistant editors was also appointed to further simplify the editing procedure and attain best results for the readers.

Apart from the editorial board, the designing team has also invested a significant amount of their time in understanding the subject and creating the most relevant covers. They scrutinized every image to scout for the most suitable representation of the subject and create an appropriate cover for the book.

The publishing team has been an ardent support to the editorial, designing and production team. Their endless efforts to recruit the best for this project, has resulted in the accomplishment of this book. They are a veteran in the field of academics and their pool of knowledge is as vast as their experience in printing. Their expertise and guidance has proved useful at every step. Their uncompromising quality standards have made this book an exceptional effort. Their encouragement from time to time has been an inspiration for everyone.

The publisher and the editorial board hope that this book will prove to be a valuable piece of knowledge for students, practitioners and scholars across the globe.

Index

A

Aerosols, 23, 38, 41, 44, 47, 49-50, 65, 67-69, 105, 142, 146, 152-156, 170, 174, 190, 199

Afforestation, 54, 208, 227

Air Pressure, 2, 14, 134

Alpine Zone, 15, 130

Annual Mean Temperature, 2, 7-8, 144

Annual Precipitation, 2, 6, 8, 11

Atmospheric Circulation, 14, 21-22, 26, 29, 38, 89, 144

B

Bathymetry, 32, 34

Biosphere, 21, 32, 51, 161-162, 201

Bond Cycle, 28

C

Carbon Cycle, 35, 47, 87, 97-98, 136, 148, 161, 167, 173

Carbon Monoxide, 39, 141, 152, 164, 169-170

Chlorofluorocarbons, 49, 56-58, 60, 169, 179, 209

Climate Feedback, 46, 90, 147, 161

Climate Sensitivity, 44-51, 124, 150, 159-160

Climate Variations, 22, 24, 166

Cloud Cover, 2, 5, 7, 66, 160-161

Continental Drift, 37

D

Decadal Variation, 20, 23

Deforestation, 37, 39, 51-55, 85, 124-125, 149, 155, 160, 165, 167, 177-178, 186, 209

Dry Seasons, 3, 6

E

Earth System Sensitivity, 45-46

Evapotranspiration, 4, 25, 31, 54, 155

G

Glacial Period, 28-29, 37, 79-81, 144

Global Dimming, 37, 64-69, 95

Greenhouse Gases, 24, 31, 34-35, 39-41, 44-45, 47, 49, 53, 59, 65, 67-68, 79, 86, 97, 108, 122, 125, 143, 147, 154, 159, 164, 167-174, 177, 188, 191, 200, 206, 210, 213, 217

Groundcover, 41

H

Heat Waves, 9, 20, 94, 106, 111, 195

Humid Tropical, 4

Hurricanes, 1, 20, 69, 87, 94, 195

Hydrosphere, 32, 148, 201

Hydroxyl Radical, 57, 150, 170, 190

I

Interannual Variation, 20, 22

Isotherms, 2

M

Marine Climate, 10

Mediterranean Climate, 10

Methane, 31, 35, 39-40, 44, 49, 52, 58, 89, 96-98, 107, 124-125, 129, 141-143, 146, 148, 150-151, 162, 165, 167-168, 170-174, 179-182, 187, 192, 200, 209, 213, 224-225

Milankovitch Cycles, 37, 70-71

Monsoon Climates, 1, 6

N

Nitrogen Oxides, 39

O

Ocean Currents, 2, 15, 18, 26, 28, 37-39, 87, 118, 127, 141, 143

Oceanic Circulation, 21-22, 33

Orbital Forcing, 27, 30, 32, 37, 50, 71, 79, 81

Orbital Variation, 19, 26

Ozone Cycle, 56-57

Ozone Depletion, 37, 55-56, 58-64, 89, 155, 175, 180

Ozone Hole, 55-56, 58-62

P

Paleocene-eocene Thermal Maximum, 32, 50, 88, 151

Polar Climate, 13

Polar Stratospheric Clouds, 59, 61

R

Radiative Equilibrium, 40, 44-45

Radiative Forcing, 37, 40-47, 49-50, 72, 145-147, 150-156, 158-160, 165, 170, 172, 174-175, 190-191

Rain Shadow Effect, 15
Relative Humidity, 9, 47, 51, 160, 182, 195
Rock Weathering, 35

S

Savanna, 4, 6
Sea Level, 2, 14, 29-30, 86-87, 94, 98-99, 101-103, 105-106, 121, 133, 136-137, 142, 162-163, 178, 191-194
Seasonal Variation, 21, 75
Soil Carbon, 52, 91, 121
Solar Energy, 41-42, 65-66, 79, 81, 154, 163
Solar Forcing, 42-43, 72, 78, 82, 157
Solar Radiation, 15, 18, 21, 26-27, 35, 38, 41, 61, 66, 69, 71-75, 79, 90, 134, 144-146, 152-155, 158, 161, 163-164, 167, 174, 199
Steppe, 4, 8
Stratosphere, 38, 44, 56-62, 69-70, 82, 84, 96-97, 150, 155-158, 161, 170-171, 182, 190, 200
Stratospheric Ozone, 55-56, 60, 62-63, 122, 155, 157, 170
Subpolar Climate, 12, 15
Subtropical Climates, 8-10
Sulphur Dioxide, 37-38
Summer Solstice, 21, 71, 73, 79

T

Tectonic Forces, 32, 34
Temperate Climates, 11
Temperate Latitudes, 5, 100
Thermal Maxima, 26
Thunderstorms, 9, 97
Transient Climate Response, 45-46, 49-50
Tropopause, 41, 97
Tropospheric Ozone, 56, 62-63, 170, 174

V

Volcanic Eruption, 38, 50, 151, 156
Volcanic Forcing, 47, 82
Volcanism, 23, 91, 105

W

Water Cycle, 68, 86, 132-133
Water Vapour, 31, 37, 39, 47, 142-143, 147, 150, 152, 155, 159-160
Windward Side, 15
Winter Solstice, 21, 73